Lecture Notes in Mathematics

Edited by A. Dold and B. Eckmann

Series: Mathematisches Institut der Universität Erlangen-Nürnberg
Advisers: H. Bauer and K. Jakobs

803

Fumi-Yuki Maeda

T0220360

Dirichlet Integrals on Harmonic Spaces

Springer-Verlag
Berlin Heidelberg New York 1980

Author

Fumi-Yuki Maeda
Dept. of Mathematics, Faculty of Science
Hiroshima University
Hiroshima, 730/Japan

AMS Subject Classifications (1980): 31 D 05

ISBN 3-540-09995-6 Springer-Verlag Berlin Heidelberg New York
ISBN 0-387-09995-6 Springer-Verlag New York Heidelberg Berlin

This work is subject to copyright. All rights are reserved, whether the whole or part of the material is concerned, specifically those of translation, reprinting, re-use of illustrations, broadcasting, reproduction by photocopying machine or similar means, and storage in data banks. Under § 54 of the German Copyright Law where copies are made for other than private use, a fee is payable to the publisher, the amount of the fee to be determined by agreement with the publisher.

© by Springer-Verlag Berlin Heidelberg 1980
Printed in Germany

Printing and binding: Beltz Offsetdruck, Hemsbach/Bergstr.
2141/3140-543210

ACKNOWLEDGEMENTS

These lecture notes were prepared for the course "Topics in Axiomatic Potential Theory" which was given at Mathematisches Institut der Universität Erlangen-Nürnberg during the academic year 1978-1979.

The author is deeply indebted to Professor H.Bauer for his invitation to the Institute which provided the author an opportunity to give lectures on the present subject. The author would like to express his gratitude to those colleagues and students in the Institute who patiently followed the lectures. Thanks are also due to Mrs. Ch.Rische of the Institute for her elaboration on the typing of the manuscript.

Fumi-Yuki MAEDA

February 1980

CONTENTS

INTRODUCTION

The classical potential theory is, in a sense, a study of the Laplace equation $\Delta u=0$. It has been clarified that second order elliptic, and some parabolic, partial differential equations share many potential theoretic properties with the Laplace equation. An axiomatic potential theory tries to develop a unified method of treating these equations.

In an axiomatic potential theory, we start with defining a harmonic space $(X, \mathcal{H})$ or $(X, \mathcal{U})$, where X is locally compact Hausdorff space and $\mathcal{H}$ (resp. $\mathcal{U}$) is a sheaf of linear spaces of continuous functions (resp. convex cones of lower semicontinuous functions) which are called "harmonic" (resp. "hyperharmonic"). There are several different kinds of harmonic spaces so far introduced. Among them, the following three are the most well-established:

(a) Brelot's harmonic space $(X, \mathcal{H})$ (see [6], [7], [16], etc.);
(b) Harmonic spaces $(X, \mathcal{H})$ given in Bauer [1] and in Boboc-Constantinescu-Cornea [2];
(c) Harmonic space $(X, \mathcal{U})$ proposed in Constantinescu-Cornea [11].

On any of these harmonic spaces, we can naturally develop a theory of superharmonic functions and potentials, including the Perron-Wiener's method for Dirichlet problems, balayage theory and even integral representation of potentials; and thus a fairly large part of the classical potential theory is covered also by axiomatic theory.

There are, however, some important parts in the classical potential theory which involve the notion of Dirichlet integrals. Due to the fact that only topological notions and some order relations are involved in an axiomatic potential theory, it is impossible to define differentiation of functions without further structures on X. However, it appears that with some reasonable additional structure for $\mathcal{H}$ or $\mathcal{U}$, we can consider a notion corresponding to the gradient of functions on a harmonic space.

As an illustration, let us consider the case where X is an euclidean domain and the harmonic sheaf $\mathcal{H}$ is given by the solutions of the second order differential equation

$$Lu \equiv \Sigma\, a_{ij} \frac{\partial^2 u}{\partial x_i \partial x_j} + \Sigma\, b_i \frac{\partial u}{\partial x_i} + cu = 0,$$

where a_{ij}, b_i, c are functions on X with certain regularity and (a_{ij}) is positive definite everywhere on X. Now, we have the following equality:

$$2\, \Sigma\, a_{ij} \frac{\partial f}{\partial x_i} \frac{\partial g}{\partial x_j} = L(fg) - fLg - gLf + fgL1.$$

This shows that the function $\Sigma\, a_{ij} \frac{\partial f}{\partial x_i} \frac{\partial g}{\partial x_j}$ (which, by an abuse of terminology, we call mutual gradient of f and g) can be expressed in terms of L. Therefore, in an axiomatic theory, once a notion corresponding to the operator L is introduced, then mutual gradients of functions can be defined by the above equation.

The purpose of the present lectures is to define the notion of (mutual) gradients of functions on harmonic spaces following the above idea, to show that this notion enjoys some basic properties possessed by the form $\Sigma\, a_{ij} \frac{\partial f}{\partial x_i} \frac{\partial g}{\partial x_j}$ and to develop some theories involving the notion of Dirichlet integrals in the axiomatic setting.

As a matter of fact, we define the gradients of functions as measures, which we call gradient measures. The definition and the verification of basic properties of gradient measures can be carried out on general harmonic spaces in the sense of Constantinescu-Cornea [11]. Thus, in Part I, we give a theory on general harmonic spaces. Sections §1 and §2 are preparatory and almost all materials in these sections are taken from Part I of [11]. In §3, we give the definition of gradient measures and prove basic properties. This section is nearly identical with [26].

In order to obtain richer results, it becomes necessary for us to restrict ourselves to self-adjoint harmonic spaces. Self-adjointness of a harmonic space is defined by the existence of consistent system of symmetric Green functions (see §4 for details); its prototype is the space given by solutions of the equation of the form $\Delta u = cu$ (c: a function). Thus, in Part II and Part III, we develop our theory on self-adjoint harmonic spaces. The main theme of Part II is Green's formula. In §4, we study Green potentials and in §5 we establish

Green's formula for a harmonic function and a potential both with finite energy. Most of the materials in Part II are taken from [24] (and also [22], [23]), but in these lectures arrangements and proofs are often different from those in [24] and the final form of Green's formula is improved. Part III is devoted to the study of various spaces of Dirichlet-finite or energy-finite functions. Spaces of harmonic functions are mainly discussed in §6. In §7, we consider a functional completion to define those functions which correspond to continuous BLD-functions in the classical theory (cf. [12], [5]). Finally in §8, we shall show that some part of the theory of Royden boundary (cf. [29], [10]) can be also developed in the axiomatic theory and a Neumann problem can be discussed (cf. [19], [20] for the classical case).

Presentations of these lectures are almost self-contained. The biggest exception is that we use without proof the existence of Green functions and the integral representation theorem for potentials on Brelot's harmonic spaces. For these one may refer to [16] and [11]. Some examples are given without detailed explanations. In the Appendix, networks are studied as examples of harmonic spaces.

Terminology and notation

Given a topological space X and a subset A of X, we denote by $\overline{A}$ the closure of A, $\overset{\circ}{A}$ the interior of A and ∂A the boundary of A. For two sets A,B, $A\setminus B$ means the difference set. The family of all open subsets of X is denoted by $\mathcal{O}_X$. A connected open set is called a domain. By a function, we shall always mean an extended real valued function. A continuous function will mean a finite-valued one. The set of all continuous functions on X is denoted by $\mathcal{C}(X)$, and the set of all $f\in\mathcal{C}(X)$ having compact supports in X is denoted by $\mathcal{C}_0(X)$. The support of f is denoted by Supp f. Given a set $A\subset X$ and a class $\mathcal{F}$ of functions on A, we say that $\mathcal{F}$ separates points of A if for any $x,y\in A$, $x\neq y$, there are $f,g\in\mathcal{F}$ satisfying $f(x)g(y)\neq f(y)g(x)$ (with convention $0\cdot\infty=\infty\cdot 0=0$). For two classes $\mathcal{F}_1$, $\mathcal{F}_2$ of finite valued functions, $\mathcal{F}_1-\mathcal{F}_2=\{f_1-f_2 \mid f_1\in\mathcal{F}_1, f_2\in\mathcal{F}_2\}$. For a class $\mathcal{F}$ of functions, $\mathcal{F}^+=\{f\in\mathcal{F} \mid f\geq 0\}$.

For a locally compact space X, a measure on X will mean a (signed) real Radon measure on X. The set of all measures on X is denoted by $\mathcal{M}(X)$. For $\mu\in\mathcal{M}(X)$, μ^+ and μ^- denotes the positive part and the negative part of μ, and $|\mu| = \mu^+ + \mu^-$. For $\mu \in \mathcal{M}(X)$ and $f \in \mathcal{C}(X)$, $f\mu$ is the measure defined by $(f\mu)(\varphi) = \mu(f\varphi)$ for $\varphi \in \mathcal{C}_0(X)$. Restriction of a function or a measure to a set A is denoted by $\cdot|A$.

By a <u>sheaf</u> of functions on X (resp. a sheaf of measures on X), we mean a mapping Φ defined on $\mathcal{O}_X$ satisfying the following three conditions:

(a) for any $U \in \mathcal{O}_X$, $\Phi(U)$ is a set of functions (resp. measures) on U;

(b) if $U, V \in \mathcal{O}_X$, $U \subset V$ and $\varphi \in \Phi(V)$, then $\varphi|V \in \Phi(U)$;

(c) if $(U_\iota)_{\iota \in I}$ is a subfamily of $\mathcal{O}_X$, φ is a function (resp. measure) on $\bigcup_{\iota \in I} U_\iota$ and if $\varphi|U_\iota \in \Phi(U_\iota)$ for all $\iota \in I$, then $\varphi \in \Phi(\bigcup_{\iota \in I} U_\iota)$.

The mapping $\mathcal{M}$: $U \mapsto \mathcal{M}(U)$ is a sheaf, which is called <u>the sheaf of measures</u> on X.

For a locally compact space X with a countable base, a sequence $\{U_n\}$ of relatively compact open sets U_n such that $\overline{U}_n \subset U_{n+1}$ for each n and $\bigcup U_n = X$ is called an <u>exhaustion</u> of X.

PART I THEORY ON GENERAL HARMONIC SPACES

§1. Harmonic spaces

In this section, we first give the definition of harmonic spaces in the sense of Constantinescu-Cornea [11]. Then, we shall show that Brelot's harmonic spaces and Bauer-Boboc-Constantinescu-Cornea's harmonic spaces are special cases.

Throughout, the base space X is assumed to be a locally compact (Hausdorff) space with countable base.

1-1. Definition of harmonic spaces (cf. [11])

A sheaf $\mathcal{U}$ of functions on X is called a hyperharmonic sheaf if for any $U \in \mathcal{O}_X$, $\mathcal{U}(U)$ is a convex cone of lower semicontinuous $]-\infty,+\infty]$-valued functions on U.
Given a hyperharmonic sheaf $\mathcal{U}$, we define

$$\mathcal{H}_{\mathcal{U}}(U) = \mathcal{U}(U) \cap -\mathcal{U}(U)$$

for each $U \in \mathcal{O}_X$. $\mathcal{H}_{\mathcal{U}}(U)$ is a linear space of continuous functions on U, and $\mathcal{H}_{\mathcal{U}}$ is a sheaf of functions on X, which is called the harmonic sheaf associated with $\mathcal{U}$.
An open set $U \in \mathcal{O}_X$ is called an MP-set for $\mathcal{U}$ if the following minimum principle is valid:

If $f \in \mathcal{U}(U)$, $f \geq 0$ on $U \setminus K$ for some compact set K in X and $\liminf_{x \to \xi,\, x \in U} f(x) \geq 0$ for every $\xi \in \partial U$, then $f \geq 0$ on U.

Let U be an MP-set. For a function φ on ∂U, we define

$$\overline{\mathcal{U}}^U_\varphi = \left\{ u \in \mathcal{U}(U) \;\middle|\; \begin{array}{l} u \text{ is lower bounded on } U, \\ u \geq 0 \text{ on } U \setminus K \text{ for some compact set } K \text{ in } X, \\ \liminf_{x \to \xi,\, x \in U} u(x) \geq \varphi(\xi) \text{ for every } \xi \in \partial U \end{array} \right\}$$

and $\underline{\mathcal{U}}^U_\varphi = -\overline{\mathcal{U}}^U_{-\varphi}$. Put

$$\overline{H}^U_\varphi = \inf \overline{\mathcal{U}}^U_\varphi \quad \text{and} \quad \underline{H}^U_\varphi = \sup \underline{\mathcal{U}}^U_\varphi$$

(if $\overline{\mathcal{U}}_\varphi^U = \emptyset$, then $\overline{H}_\varphi^U \equiv +\infty$; if $\underline{\mathcal{U}}_\varphi^U = \emptyset$, then $\underline{H}_\varphi^U \equiv -\infty$). Then, from the definitions, the following properties are easily seen:

$$-\overline{H}_\varphi^U = \underline{H}_{-\varphi}^U, \quad \underline{H}_\varphi^U \leq \overline{H}_\varphi^U,$$

$$\overline{H}_{\alpha\varphi} = \alpha\overline{H}_\varphi \quad \text{if } \alpha \text{ is a constant and } \alpha \geq 0,$$

$$\varphi \leq \psi \text{ on } \partial U \text{ implies } \overline{H}_\varphi^U \leq \overline{H}_\psi^U \text{ and } \underline{H}_\varphi^U \leq \underline{H}_\psi^U$$

$$\overline{H}_{\varphi+\psi}^U \leq \overline{H}_\varphi^U + \overline{H}_\psi^U, \text{ provided that } +\infty-\infty \text{ or } -\infty+\infty$$

does not occur.

A function φ on ∂U is called <u>resolutive</u> (for U, with respect to $\mathcal{U}$) if $\overline{H}_\varphi^U = \underline{H}_\varphi^U$ and it belongs to $\mathcal{H}_{\mathcal{U}}(U)$. In this case we denote $\overline{H}_\varphi^U = \underline{H}_\varphi^U$ by H_φ^U. A non-empty open set $U \in \mathcal{O}_X$ is called a <u>resolutive set</u> (with respect to $\mathcal{U}$) if it is an MP-set and every $\varphi \in \mathcal{C}_0(\partial U)$ is resolutive. If U is a resolutive set, then for each $x \in U$ the map $\varphi \mapsto H_\varphi^U(x)$ is a positive linear functional on $\mathcal{C}_0(\partial U)$. Hence, there exists a non-negative measure μ_x^U on ∂U such that

$$H_\varphi^U(x) = \int \varphi d\mu_x^U \quad \text{for all } \varphi \in \mathcal{C}_0(\partial U).$$

This measure μ_x^U is called the <u>harmonic measure</u> of U at x (with respect to $\mathcal{U}$). For a function φ on ∂U, we define $\mu^U\varphi$ by

$$(\mu^U\varphi)(x) = \int^* \varphi d\mu_x^U.$$

In particular, $\mu^U\varphi = H_\varphi^U$ if $\varphi \in \mathcal{C}_0(\partial U)$.

A pair $(X, \mathcal{U})$ of a locally compact space X (with countable base) and a hyperharmonic sheaf $\mathcal{U}$ on X is called a <u>harmonic space</u> if the following four axioms are satisfied:

(P)(Axiom of positivity): For each $x \in X$, there is $U \in \mathcal{O}_X$ and $h \in \mathcal{H}_{\mathcal{U}}(U)$ such that $x \in U$ and $h(x) \neq 0$.

(R)(Axiom of resolutivity): The resolutive sets with respect to $\mathcal{U}$ form a base of the topology of X.

(C)(Axiom of completeness): For any open set U, a lower semicontinuous $]-\infty,+\infty]$-valued function u on U belongs to $\mathcal{U}(U)$ if, for any relatively compact resolutive set V such that $\overline{V} \subset U$, $\mu^V u \leq u$ on V.

(BC) (Bauer convergence property): For any $U \in \mathcal{O}_X$, if $\{u_n\}$ is a monotone increasing sequence of functions in $\mathcal{H}_{\mathcal{U}}(U)$ and if it is locally uniformly bounded on U, then the limit function $u = \lim_{n\to\infty} u_n$ belongs to $\mathcal{H}_{\mathcal{U}}(U)$.

Remark 1.1. In Axiom (P), the condition $h(x) \neq 0$ may be replaced by $h(x) > 0$. Furthermore, by choosing U small enough, we may require $h > 0$ on U, or even on $\overline{U}$.

Remark 1.2. By (C) and the fact that $\mathcal{U}$ is a sheaf, we have the following: For $U \in \mathcal{O}_X$ and a lower semicontinuous $]-\infty,+\infty]$-valued function u on U, if every $x \in U$ has an open neighborhood V_x such that, whenever V is a relatively compact resolutive set with $\overline{V} \subset V_x$, $\mu^V u \leq u$ on V, then $u \in \mathcal{U}(U)$.

Given a harmonic space $(X, \mathcal{U})$, functions in $\mathcal{H}_{\mathcal{U}}(U)$ are called harmonic on U and functions in $\mathcal{U}(U)$ are called hyperharmonic on U. If $-u$ is hyperharmonic on U, then u is called hypoharmonic on U.

Let $Y \in \mathcal{O}_X$ $(Y \neq \emptyset)$ and let $f \in \mathcal{C}(Y)$ be strictly positive on Y. For each $U \in \mathcal{O}_Y$ put

$$\mathcal{U}_{Y,f}(U) = \{u/f \mid u \in \mathcal{U}(U)\}.$$

Then, $\mathcal{U}_{Y,f}$ is a hyperharmonic sheaf on Y and $(Y, \mathcal{U}_{Y,f})$ is a harmonic space. In case $Y=X$, $\mathcal{U}_{Y,f}$ will be denoted by $\mathcal{U}_f$; in case $f \equiv 1$, $\mathcal{U}_{Y,f}$ will be denoted by $\mathcal{U}_Y$ and $(Y, \mathcal{U}_Y)$ is called the restriction of $(X, \mathcal{U})$ to Y.

1-2. Brelot's harmonic spaces (cf. [6], [7], [11;Chap.3])

A pair $(X, \mathcal{H})$ of a locally compact space X and a sheaf $\mathcal{H}$ of functions on X is called a Brelot's harmonic space if it satisfies the following three axioms:

Axiom 1. For any $U \in \mathcal{O}_X$, $\mathcal{H}(U)$ is a linear subspace of $\mathcal{C}(U)$.

Axiom 2. Regular domains (with respect to $\mathcal{H}$) form a base of the topology of X.

Here, a domain V in X is called <u>regular</u> with respect to $\mathcal{H}$ if it is relatively compact, $\partial V \neq \phi$ and for each $\varphi \in \mathcal{C}(\partial V)$ there is a unique $u \in \mathcal{C}(\overline{V})$ such that $u|\partial V = \varphi$ and $u|V \in \mathcal{H}(V)$, and such that $\varphi \geq 0$ implies $u \geq 0$.

Axiom 3. If U is a domain in X, $\{u_n\}$ is a monotone increasing sequence of functions in $\mathcal{H}(U)$ and $\{u_n(x_o)\}$ is bounded for some $x_o \in U$, then $u = \lim_{n\to\infty} u_n$ belongs to $\mathcal{H}(U)$.

If V is a regular domain and $\varphi \in \mathcal{C}(\partial V)$, the function $u \in \mathcal{C}(\overline{V})$ satisfying $u|\partial V = \varphi$ and $u|V \in \mathcal{H}(V)$ is denoted by H_φ^U. Then the mapping $\varphi \mapsto H_\varphi^U(x)$ is positive linear on $\mathcal{C}(\partial V)$, so that the harmonic measure μ_x^V of V at $x \in V$ is defined as in the case of resolutive sets, and we define μ^V similarly.

Let $U \in \mathcal{O}_X$ and let u be a lower semicontinuous $]-\infty,+\infty]$-valued function on U. u is called locally hyperharmonic on U (with respect to $\mathcal{H}$) if every $x \in U$ has an open neighborhood V_x such that, whenever V is a regular domain with $\overline{V} \subset V_x$, $\mu^V u \leq u$ on V. Let $\mathcal{U}_{\mathcal{H}}(U)$ be the class of all locally hyperharmonic functions on U (with respect to $\mathcal{H}$). Then it is easy to see that $\mathcal{U}_{\mathcal{H}}$ is a hyperharmonic sheaf on X and $\mathcal{H}_{\mathcal{U}_{\mathcal{H}}} = \mathcal{H}$.

<u>Lemma 1.1.</u> Let $(X, \mathcal{H})$ be a Brelot's harmonic space, $U \in \mathcal{O}_X$ is a domain and $u \in \mathcal{H}(U)$. If $u \geq 0$ on U and $u(x_o) = 0$ for some $x_o \in U$, then $u = 0$.

<u>Proof.</u> Let $u_n = nu$ $(n=1,2,\ldots)$. Then $u_n \in \mathcal{H}(U)$, $\{u_n\}$ is monotone increasing and $\{u_n(x_o)\}$ is bounded. Hence, $\lim_{n\to\infty} u_n \in \mathcal{H}(U)$ by Axiom 3, which implies $u(x) = 0$ for all $x \in U$.

<u>Lemma 1.2.</u> Let $(X, \mathcal{H})$ be a Brelot's harmonic space, V be a regular domain and W be an open set such that $\partial V \cap W \neq \emptyset$. Then $\mu_x^V(\partial V \cap W) > 0$ for all $x \in V$.

<u>Proof.</u> Choose $\varphi \in \mathcal{C}(\partial V)$ such that $0 \leq \varphi \leq 1$ on ∂V, Supp $\varphi \subset \partial V \cap W$ and $\varphi \neq 0$. By the above lemma, $H_\varphi^V(x) > 0$ for all $x \in V$. Hence, $\mu_x^V(\partial V \cap W) \geq H_\varphi^V(x) > 0$.

Lemma 1.3. Let $(X, \mathcal{H})$ be a Brelot's harmonic space, U be a domain in X and $u \in \mathcal{U}_{\mathcal{H}}(U)$. If $u \equiv +\infty$ on a non-empty open set $W \subset U$, then $u \equiv +\infty$ on U.

Proof. Let $U' = \{x \in U \mid u \equiv +\infty$ on a neighborhoof of $x\}$. Then U' is non-empty and open. Suppose $U' \neq U$. Let U_1 be a connected component of U'. Since U is connected, $\partial U' \cap U \neq \emptyset$. Let $x_1 \in \partial U_1 \cap U$. Choose an open set V_1 such that $x_1 \in V_1$ and $\mu^V u \leq u$ for all regular domain V with $\overline{V} \subset V_1$. Choose $y_1 \in V_1 \cap U_1$ and choose a regular domain V such that $x_1 \in V$ and $\overline{V} \subset V_1 \setminus \{y\}$. Since U_1 is connected, $\partial V \cap U_1 \neq \emptyset$. Since $u \equiv +\infty$ on $\partial V \cap U_1$, the previous lemma implies $u(x) \geq \mu^V u(x) = +\infty$ for all $x \in V$. Therefore, $x_1 \in U'$, which is a contradiction. Thus $U' = U$, and the lemma is proved.

Proposition 1.1. Let $(X, \mathcal{H})$ be a Brelot's harmonic space, U be a domain in X, $u \in \mathcal{U}_{\mathcal{H}}(U)$ and $u \geq 0$ on U. If $u(x_o) = 0$ for some $x_o \in U$, then $u = 0$.

Proof. The set $U^+ = \{x \in U \mid u(x) > 0\}$ is open. Suppose $U^+ \neq \emptyset$. Let $u_\infty = \lim_{n\to\infty} nu$. Clearly, $u_\infty \in \mathcal{U}_{\mathcal{H}}(U)$. Since $u_\infty \equiv +\infty$ on U_1, the previous lemma implies $u_\infty \equiv +\infty$, i.e., $u(x) > 0$ for all $x \in U$.

Proposition 1.2. (Minimum principle) Let $(X, \mathcal{H})$ be a Brelot's harmonic space, $U \in \mathcal{O}_X$ and suppose there is $u_o \in \mathcal{U}_{\mathcal{H}}(U) \cap \mathcal{C}(U)$ such that $\inf_U u_o > 0$. Then U is an MP-set with respect to $\mathcal{U}_{\mathcal{H}}$.

Proof: Let $u \in \mathcal{U}_{\mathcal{H}}(U)$ and suppose $u \geq 0$ on $U \setminus K$ for some compact set K in X and $\liminf_{x\to\xi, x\in U} u(x) \geq 0$ for every $\xi \in \partial U$. Put $\alpha \equiv \inf_U(u/u_o)$. Suppose $\alpha < 0$. Then, by the lower semicontinuity of u/u_o and the boundary condition for u, we see that there is $x_o \in U$ such that $\alpha = u(x_o)/u_o(x_o)$. The function $v = u - \alpha u_o$ belongs to $\mathcal{U}_{\mathcal{H}}(U)$, $v \geq 0$ on U and $v(x_o) = 0$. Hence, by Proposition 1.1, $u = \alpha u_o$ on the component U' of U which contains x_o. Since $\alpha < 0$, this fact contradicts our boundary condition for u.

<u>Theorem 1.1.</u> If $(X, \mathcal{H})$ is a Brelot's harmonic space, then $(X, \mathcal{U}_{\mathcal{H}})$ is a harmonic space (in the sense of [11]). Furthermore, any locally hyperharmonic functions are hyperharmonic, i.e., if $u \in \mathcal{U}_{\mathcal{H}}(U)$, then $\mu^V u \leq u$ on V for any regular domain V such that $\overline{V} \subset U$.

<u>Proof.</u> Let V be a regular domain and let $u_o = H_1^V$. Then $u_o \in \mathcal{H}(V)$ and $u_o > 0$ on $\overline{V}$ by Lemma 1.1. Hence, by Axiom 2, Axiom (P) is satisfied. Furthermore, by Proposition 2, we see that any regular domain is an MP-set. Then it is easy to see that $H_\varphi^V = \overline{H}_\varphi^V = \underline{H}_\varphi^V$ for $\varphi \in \mathcal{C}(\partial V)$ for a regular domain V. Thus, a regular domain is resolutive with respect to $\mathcal{U}_{\mathcal{H}}$, and hence Axiom 2 implies Axiom (R). Axiom (C) is an immediate consequence of the definition of $\mathcal{U}_{\mathcal{H}}$ and Axiom (BC) is a weaker form of Axiom 3. The last assertion of the theorem follows from the fact that every regular domain is an MP-set with respect to $\mathcal{U}_{\mathcal{H}}$.

<u>1-3. Bauer-Boboc-Constantinescu-Cornea's harmonic space</u> (cf. [1], [2], [11,Chap.3])

Let X be a locally compact space and $\mathcal{H}$ a sheaf of functions on X satisfying Axioms 1 and 2 of Brelot. As in the case of Brelot's harmonic space, let $\mathcal{U}_{\mathcal{H}}$ be the sheaf of locally hyperharmonic functions with respect to $\mathcal{H}$. Let $\mathcal{H}^*(U)$ be the set of all hyperharmonic functions on U, i.e.,

$$\mathcal{H}^*(U) = \left\{ u \;\middle|\; \begin{array}{l} \text{lower semicontinuous }]-\infty,+\infty]\text{-valued}, \\ \mu^V u \leq u \text{ for } \underline{\text{all}} \text{ regular domain } V \text{ with } \overline{V} \subset U \end{array} \right\}.$$

The pair $(X, \mathcal{H})$ is called a Bauer-Boboc-Constantinescu-Cornea's (or, simply, Bauer's; cf. [11; Chap.3]) harmonic space if, in addition to Axioms 1 and 2 of Brelot, it satisfies Axioms (P), (BC) and the following (S):

(S) For any $x \in X$, there is an open neighborhood V of x for which $\mathcal{H}^*(V)$ separates points of V.

It can be shown that Brelot's harmonic space is a Bauer's harmonic space; we postpone its proof to §2 (Remark 2.2). Here, we shall show that Bauer's harmonic space is a harmonic space (in the sense of [11]).

Lemma 1.4. (Bauer) Let Y be a compact set and $\mathcal{F}$ be a family of lower semicontinuous $]-\infty,+\infty]$-valued functions on Y. Suppose $\mathcal{F}$ separates points of Y and there is $g\in\mathcal{F}$ which is continuous and strictly positive on Y. If $f\in\mathcal{F}$ and $f(x) < 0$ for some $x\in Y$, then there exists $x_o\in Y$ such that $f(x_o) < 0$ and the unit point mass ε_{x_o} at x_o is the only non-negative measure μ on Y satisfying

$$\int u\, d\mu \leq u(x_o) \qquad \text{for all } u\in\mathcal{F}.$$

Proof. Put $\alpha = -\inf_Y (f/g)$. Then $\alpha > 0$ and $f+\alpha g \geq 0$ on Y. Since f/g is lower semicontinuous on the compact set Y,

$$K = \{y\in Y \mid f(y) + \alpha g(y) = 0\}$$

is non-empty. Obviously, K is a compact set and $f < 0$ on K. For each $y\in Y$ put

$$\mathcal{M}_y = \{\mu\in \mathcal{M}^+(Y) \mid \int u\, d\mu \leq u(y) \quad \text{for all } u\in\mathcal{F}\}$$

and

$$\mathcal{O} = \left\{ A \subset Y \;\middle|\; \begin{array}{l} A \neq \phi, \text{ compact,} \\ \text{if } y\in A \text{ and } \mu\in\mathcal{M}_y \text{ then } \mu(Y\setminus A) = 0 \end{array} \right\}.$$

If $y\in K$ and $\mu\in\mathcal{M}_y$, then $0 \leq \int (f + \alpha g)d\mu \leq (f + \alpha g)(y) = 0$, so that $f + \alpha g = 0$ μ-a.e., i.e., $\mu(Y\setminus K)=0$. Hence $K\in\mathcal{O}$. If we consider the converse inclusion relation in $\mathcal{O}$, then it is inductive; in fact if $\mathcal{L}\subset\mathcal{O}$ is linearly ordered, then $A_o = \cap\mathcal{L}$ belongs to $\mathcal{O}$. Therefore, by Zorn's lemma, there is a minimal set $\tilde{A}$ in $\mathcal{O}$ which is contained in K. We shall show that $\tilde{A}$ consists of a single point. Let $u\in\mathcal{F}$ and $u \not\equiv +\infty$ on $\tilde{A}$. Then $(u + \beta f)(x') < 0$ for some $\beta > 0$ and $x'\in\tilde{A}$. Let $\gamma = -\inf_A[(u + \beta f)/g]$. Then $\gamma > 0$, $u + \beta f + g\gamma \geq 0$ on $\tilde{A}$ and $A' = \{y\in\tilde{A} \mid u(y) + \beta f(y) + \gamma g(y) = 0\}$ is non-empty. By the same argument as for K, we see that $A'\in\mathcal{O}$. Since $\tilde{A}$ is minimal, $A' = \tilde{A}$, which means that $u = -\beta f - \gamma g = (\alpha\beta - \gamma)g$ on $\tilde{A}$. Thus, every $u\in\mathcal{F}$ is proportional to g on $\tilde{A}$. Since $\mathcal{F}$ separates points of Y, it follows that $\tilde{A}$ consists of a single point: $\tilde{A} = \{x_o\}$. If $\mu\in\mathcal{M}_{x_o}$ then $\mu(Y\setminus\{x_o\}) = 0$, so that $\mu = c\varepsilon_{x_o}$ for some $c \geq 0$. Since $\int u\, d\mu \leq u(x_o)$ for all $u\in\mathcal{F}$ and $f(x_o) < 0$, $g(x_o) > 0$, we see that $c = 1$. Therefore, $\mu = \varepsilon_{x_o}$, i.e., $\mathcal{M}_{x_o} = \{\varepsilon_{x_o}\}$.

<u>Proposition 1.3.</u> (Minimum principle) Let $(X, \mathcal{H})$ be a Bauer's harmonic space and $U \in \mathcal{O}_X$. Suppose

(a) there exists $v \in \mathcal{U}_{\mathcal{H}}(U) \cap \mathcal{C}(U)$ such that $\inf_U v > 0$, and

(b) $\mathcal{H}^*(U)$ separates points of U.

Then U is an MP-set with respect to $\mathcal{U}_{\mathcal{H}}$ and $\mathcal{U}_{\mathcal{H}}(U) = \mathcal{H}^*(U)$.

<u>Proof.</u> Let $u \in \mathcal{U}_{\mathcal{H}}(U)$, $u \geq 0$ on $U \setminus K$ for some compact set K in X and $\liminf_{x \to \xi, x \in U} u(x) \geq 0$ for all $\xi \in \partial U$. Put

$$A = \{x \in U \mid u(x) < 0\}$$

and suppose $A \neq \emptyset$. Choose $\varepsilon > 0$ such that $u(x) + \varepsilon\, v(x) < 0$ for some $x \in A$ and put

$$K' = \{y \in A \mid u(y) + \varepsilon v(y) \leq 0\}.$$

Then, K' is a non-empty compact set. Let Y be a compact neighborhood of K' contained in U and consider the family

$$\mathcal{F} = \{u|Y \mid u \in \mathcal{H}^*(U)\} \cup \{(u+\varepsilon v)|Y\} \cup \{v|Y\}.$$

By the previous lemma, there is $x_0 \in K'$ such that ε_{x_0} is the only non-negative measure μ on Y satisfying $\int f\, d\mu \leq f(x_0)$ for all $f \in \mathcal{F}$. Choose a regular domain V such that $x_0 \in V$, $\overline{V} \subset Y$, $\int u\, d\mu_{x_0}^V \leq u(x_0)$ and $\int v\, d\mu_{x_0}^V \leq v(x_0)$. Then $\int f\, d\mu_{x_0}^V \leq f(x_0)$ for all $f \in \mathcal{F}$, and hence $\mu_{x_0}^V = \varepsilon_{x_0}$, which is absurd. Therefore, $A = \emptyset$, i.e., $u \geq 0$ on U.

The assertion $\mathcal{U}_{\mathcal{H}}(U) = \mathcal{H}^*(U)$ now follows from the fact that any regular domain V with $\overline{V} \subset U$ is an MP-set by the above result.

Theorem 1.2. If $(X, \mathcal{H})$ is a Bauer's harmonic space, then $(X, \mathcal{U}_{\mathcal{H}})$ is a harmonic space (in the sense of [11]).

Proof. Axioms (P) and (BC) are included in the definition of Bauer's harmonic space. Any regular domain V for which there is $h \in \mathcal{H}(V)$ with $\inf_V h > 0$ and for which $\mathcal{H}^*(V)$ separates points of V is an MP-set by Proposition 1.3, so that it is resolutive. Hence, in view of Axioms 2, (P) and (S), we have Axiom (R). Axiom (C) is an immediate consequence of the definition of $\mathcal{U}_{\mathcal{H}}$.

Remark 1.3. Conversely, we can show that if $(X, \mathcal{U})$ is a harmonic space and if $\mathcal{H}_{\mathcal{U}}$ satisfies Axiom 2, then $(X, \mathcal{H}_{\mathcal{U}})$ is a Bauer's harmonic space; see [11; Corollary 3.1.2]. This fact also implies that a Brelot's harmonic space is a Bauer's harmonic space.

1-4. Examples

Example 1.1. Let X be an open set in the Euclidean space $\mathbb{R}^n (n \geq 1)$. Consider an elliptic linear differential operator

$$Lu = \sum_{i,j=1}^{n} a_{ij} \frac{\partial^2 u}{\partial x_i \partial x_i} + \sum_{i=1}^{n} b_i \frac{\partial u}{\partial x_i} + cu,$$

where a_{ij}, b_i, c are locally Hölder continuous on X and (a_{ij}) is symmetric positive definite on X. For $U \in \mathcal{O}_X$, let

$$\mathcal{H}(U) = \{u \in \mathcal{C}^2(U) \mid Lu = 0 \text{ on } U\}.$$

Then, it is known (cf. [16] or [11; Exercise 3.2.7]; also cf. [21]) that $(X, \mathcal{H})$ is a Brelot's harmonic space.

In particular, if $L = \Delta$, the Laplacian, then $\mathcal{H}$ is the sheaf of classical harmonic functions; in this case any ball B such that $\overline{B} \subset X$ is regular by virtue of Poisson's integral, and Axiom 3 is known as Harnack's principle.

Example 1.2. Let X be an open set in $\mathbb{R}^{n+1}$ $(n\geq 1)$ and for $U\in \mathcal{O}_X$, let

$$\mathcal{H}(U) = \{u\in \mathcal{C}^2(U) \mid \frac{\partial u}{\partial x_{n+1}} = \Delta_n u\},$$

where $\Delta_n = \Sigma_{i=1}^{n} \frac{\partial^2 u}{\partial x_i^{\,2}}$. Then, it is known that $(X, \mathcal{H})$ is a Bauer's harmonic space (cf. [1], [11; §3.3]), but is not a Brelot's harmonic space.

Other examples of Bauer's harmonic spaces which are not Brelot's can be provided by networks. A general discussion on networks will be given in the Appendix. Here, we give the simplest case.

Example 1.3. Let $X = \mathbb{R}$ and for an open interval $U\in \mathcal{O}_X$ let

$$\mathcal{H}(U) = \begin{cases} \{u(t) = at + b \mid a,b\in R\} & \text{if } 0\notin U \\ \{u(t) = \begin{cases} b, & t\geq 0 \\ at+b, & t\leq 0 \end{cases} \mid a,b\in R\} & \text{if } 0\in U \end{cases}$$

and for a general $U\in \mathcal{O}_X$, let $\mathcal{H}(U) = \{u \mid u|U'\in \mathcal{H}(U')$ for any component U' of $U\}$. Then $\mathcal{H}$ is a sheaf on X satisfying Axiom 1. It is easy to see that any bounded open interval is regular, so that Axiom 2 is satisfied. Also, Axioms (P) and (BC) are easily verified. To see Axiom (S), if $x \neq 0$, then for any open interval V such that $0\notin V$ and $x\in V$, $\mathcal{H}(V)$ already separates points of V; if $x = 0$, then for $V =]-1,1[$, functions of the form

$$v(t) = \begin{cases} at+b, & t\geq 0 \\ a't+b, & t\leq 0 \end{cases}, \qquad a\leq 0,\ a',b\in \mathbb{R}$$

belong to $\mathcal{H}^*(V)$ and they separate points of V. Thus, $(X, \mathcal{H})$ is a Bauer's harmonic space. If we consider

$$u_n(t) = \begin{cases} 0 & , \ t\geq 0 \\ -nt & , \ t\leq 0 \end{cases}, \qquad n=1,2,\dots,$$

then we see that Axiom 3 is not satisfied, and hence $(X, \mathcal{H})$ is not a Brelot's harmonic space.

Example 1.4. ([11; Theorem 2.1.2]) Let $X = \mathbb{R}$ and for $U \in \mathcal{O}_X$, let $\mathcal{U}(U)$ be the set of all lower semicontinuous $]-\infty,+\infty]$-valued functions on U which are monotone decreasing on every component of U. Then $(X, \mathcal{U})$ is a harmonic space, but no open set is regular with respect to $\mathcal{U}$. Thus, this harmonic space is not a Bauer's harmonic space.

1-5. Properties of the base space of a harmonic space (cf. [11])

Now, we return to general harmonic spaces $(X, \mathcal{U})$ as defined in 1-1.

Lemma 1.5. (Cornea) Let Y be a locally compact space and $\mathcal{F}$ be an upper directed family of functions in $\mathcal{C}(Y)$ such that, for any increasing sequence $\{f_n\}$ in $\mathcal{F}$, $\lim_{n\to\infty} f_n$ is continuous. Then $\sup\mathcal{F}$ is continuous.

Proof. Let $f = \sup \mathcal{F}$. Suppose f is not continuous at $x_o \in Y$. Since f is lower semicontinuous, $f(x_o) < \limsup_{x\to x_o} f(x)$. Choose real numbers α and β such that

$$f(x_o) < \alpha < \beta < \limsup_{x\to x_o} f(x).$$

Choose any $f_1 \in \mathcal{F}$. Then $f_1(x_o) < \alpha$, so that there is a compact neighborhood K_1 of x_o such that $f_1 < \alpha$ on K_1. Choose $x_1 \in K_1$ such that $f(x_1) > \beta$. By induction, we can choose $f_n \in \mathcal{F}$, compact neighborhoods K_n of x_o and points $x_n \in K_n$ such that $K_n \supset K_{n+1}$, $f_n \leq f_{n+1}$, $f_n < \alpha$ on K_n and $f_{n+1}(x_n) > \beta$, $n = 1,2,\ldots$. By assumption, $g = \lim_{n\to\infty} f_n$ is continuous. $\{x_n\}$ has a limit point $x' \in \bigcap_{n=1}^{\infty} K_n$. Obviously $g(x') \leq \alpha$, since $f_n < \alpha$ on K_n. On the other hand, $g(x_n) \geq f_{n+1}(x_n) > \beta$ for all n, which implies that $g(x') \geq \beta$, a contradiction.

Proposition 1.4. Let $U \in \mathcal{O}_X$ and let $\mathcal{V}$ be an upper directed family of functions in $\mathcal{H}_{\mathcal{U}}(U)$. If $\mathcal{V}$ is locally uniformly bounded on U, then $\sup \mathcal{V} \in \mathcal{H}_{\mathcal{U}}(U)$.

Proof. By Axiom (BC) and by the above lemma, $u = \sup \mathcal{V}$ is continuous. Let V be any relatively compact open set such that $\overline{V} \subset U$. By Dini's method we can choose an increasing sequence $\{u_n\}$ in $\mathcal{V}$ which converges (uniformly) to u on V. By (BC) again, we see that $u|V \in \mathcal{H}_{\mathcal{U}}(V)$. Since $\mathcal{H}_{\mathcal{U}}$ is a sheaf, it follows that $u \in \mathcal{H}_{\mathcal{U}}(U)$.

Remark 1.4. We may as well use Choquet's lemma to prove Proposition 1.4 (cf. [6]).

Theorem 1.3. If $(X, \mathcal{U})$ is a harmonic space, then X is locally connected.

Proof. Let $x \in X$. Choose an open set U containing x such that there is $h \in \mathcal{H}_{\mathcal{U}}(U)$ with $h > 0$ on U (see Remark 1.1). Put

$$\mathcal{V} = \{V \in \mathcal{O}_X \mid x \in V \subset U,\ V \text{ is relatively closed in } U\}.$$

For each $V \in \mathcal{V}$ let

$$h_V = \begin{cases} 0 & \text{on } V \\ h & \text{on } U \setminus V. \end{cases}$$

Then $h_V \in \mathcal{H}_{\mathcal{U}}(U)$. If $V_1, V_2 \in \mathcal{V}$, then $V_1 \cap V_2 \in \mathcal{V}$ and

$$h_{V_1 \cap V_2} = \sup(h_{V_1}, h_{V_2}).$$

Hence, $\{h_V \mid V \in \mathcal{V}\}$ is upper directed. Since $0 \leq h_V \leq h$, this family is locally uniformly bounded on U. Thus, by the previous proposition, $g = \sup_{V \in \mathcal{V}} h_V$ belongs to $\mathcal{H}_{\mathcal{U}}(U)$, so that g is continuous on U. Then the set

$$V_o = \{y \in U \mid g(y) = 0\} = \{y \in U \mid g(y) < h(y)\}$$

belongs to $\mathcal{V}$, and hence it is minimal. It follows that V_o is a connected neighborhood of x.

Corollary 1.1. A harmonic space possesses a base consisting of resolutive domains.

Proof. It is easy to show that any connected component of a resolutive set is again resolutive. Hence, by this theorem and Axiom (R), we obtain the corollary.

Proposition 1.5. If $(X, \mathcal{U})$ is a harmonic space, then X has no isolated point.

Proof. If $x \in X$ is isolated, then, by Axiom (R), $\{x\}$ is resolutive, in particular, an MP-set. Therefore, $u \geq 0$ for all $u \in \mathcal{U}(\{x\})$, which implies $\mathcal{H}_{\mathcal{U}}(\{x\}) = \{0\}$, contradicting Axiom (P).

1-6. Properties of hyperharmonic functions (cf. [11])

In the rest of this section, we fix a harmonic space $(X, \mathcal{U})$.

Proposition 1.6. (a) If $u_1, u_2 \in \mathcal{U}(U)$, then $\min(u_1, u_2) \in \mathcal{U}(U)$.

(b) If $\{u_\iota\}$ is an upper directed familiy in $\mathcal{U}(U)$, then $\sup_\iota u_\iota \in \mathcal{U}(U)$.

Proof. Both assertions are easy consequences of Axiom (C).

Proposition 1.7. Let $U, U' \in \mathcal{O}_X$, $U' \subset U$ and let $u \in \mathcal{U}(U)$, $v \in \mathcal{U}(U')$. If the function

$$u^* = \begin{cases} \min(u,v) & \text{on } U' \\ u & \text{on } U \setminus U' \end{cases}$$

is lower semicontinuous on U, then $u^* \in \mathcal{U}(U)$.

Proof. Let V be a relatively compact resolutive set such that $\overline{V} \subset U$ and there is $h \in \mathcal{H}_{\mathcal{U}}(V)$ with $h > 0$ on V. Given $\varphi \in \mathcal{C}(\partial V)$ with $\varphi \leq u^*$ on ∂V, we shall show that $\mu^V\varphi \leq u^*$. Since $u^* \leq u$ on U, $\mu^V\varphi \leq u$ on V. Let $s \in \underline{\mathcal{U}}_\varphi^V$ and $\varepsilon > 0$.

Since

$$\liminf_{x \to \xi, x \in U' \cap V} \{u^*(x) + \varepsilon h(x) - s(x)\} \geq u(\xi) + \varepsilon h(\xi) - (\mu^V\varphi)(\xi)$$

$$\geq \varepsilon h(\xi) > 0$$

for every $\xi \in \partial U' \cap V$, $\min(u^* + \varepsilon h - s, 0) = 0$ on a neighborhood of $\partial U' \cap V$ in $U' \cap V$. Hence, if we put

$$w_\varepsilon = \begin{cases} \min(u^* + \varepsilon h - s, 0) & \text{on } U' \cap V \\ 0 & \text{on } V \setminus U', \end{cases}$$

then $w_\varepsilon \in \mathcal{U}(V)$. Furthermore,

$$\liminf_{x \to \eta, x \in V} w_\varepsilon(x) \geq u^*(\eta) - \varphi(\eta) \geq 0$$

for $\eta \in \partial V \cap U'$, so that $\liminf_{x \to \eta, x \in V} w_\varepsilon(x) \geq 0$ for all $\eta \in \partial V$. Since V is an MP-set, it follows that $w_\varepsilon \geq 0$. Therefore,

$u^* + \varepsilon h \geq s$ on $U' \cap V$. Letting $\varepsilon \to 0$ and taking the supremum of s, we have $u^* \geq \mu^V \varphi$ on $U' \cap V$. On the other hand, $u^* = u \geq \mu^V \varphi$ on $V \setminus U'$. Thus $u^* \geq \mu^V \varphi$.

Now, we conclude that $u^* \in \mathcal{U}(U)$ by Axiom (C) (cf. Remark 1.2).

Proposition 1.8. Any open subset of an MP-set is again an MP-set.

Proof. Let U be an MP-set and U' be an open subset of U. Let $u \in \mathcal{U}(U')$ satisfy $u \geq 0$ on $U' \setminus K$ for some compact set K in X and $\liminf_{x \to \xi, x \in U'} u(x) \geq 0$ for all $\xi \in \partial U'$. Consider the function

$$u^* = \begin{cases} \inf(u,0) & \text{on } U' \\ 0 & \text{on } U \setminus U'. \end{cases}$$

Then, u^* is lower semicontinuous. Hence, by the previous proposition $u^* \in \mathcal{U}(U)$. Since U is an MP-set, it follows that $u^* \geq 0$, i.e., $u \geq 0$ on U'. Therefore, U' is an MP-set.

Proposition 1.9. Let $U \in \mathcal{O}_X$ and V be a relatively compact resolutive set such that $\overline{V} \subset U$. Let $u \in \mathcal{U}(U)$, $v \in \mathcal{U}(V)$ and suppose $v \geq \mu^V u$ on V. Put

$$u^*(y) = \begin{cases} u(y) & \text{if } y \in U \setminus \overline{V} \\ \min(u(y), v(y)) & \text{if } y \in V \\ \min\{u(y), \liminf_{x \to y, x \in V} v(x)\} & \text{if } y \in \partial V. \end{cases}$$

If $u^*(y) > -\infty$ for all $y \in \partial V$, then $u^* \in \mathcal{U}(U)$.

Proof. Obviously, u^* is lower semicontinuous on U. Let W be any relatively compact resolutive set such that $\overline{W} \subset U$ and consider $\varphi \in \mathcal{C}(\partial W)$ satisfying $\varphi \leq u^*$ on ∂W. We shall show that $u^* \geq \mu^W \varphi$. Then we see that $u^* \in \mathcal{U}(U)$.
Since $u^* \leq u$, $\mu^W \varphi \leq u$ on W, so that $\mu^W \varphi \leq u^*$ on $U \setminus \overline{V}$.
Next, let $w \in \underline{\mathcal{U}}^W_\varphi$ and extend it to $\overline{W}$ by

$$w(y) = \limsup_{x \to y, x \in W} w(x), \qquad y \in \partial W.$$

Then, $w \leq \varphi \leq u^* \leq u$ on ∂W. Therefore, $w \leq u$ on $\overline{W}$. Since w is upper semicontinuous and u is lower semicontinuous, we can choose $\psi \in \mathcal{C}(\partial V)$ such that $\psi \leq u$ on ∂V and $w \leq \psi$ on $\overline{W} \cap \partial V$. Take any $s \in \bar{\mathcal{U}}_{\psi}^{V}$ and consider the function

$$f = v + s - w - \mu^V\psi$$

on V. Obviously, f is hyperharmonic on $V \cap W$. If $y \in \partial W \cap \overline{V}$, then

$$\liminf_{x\to y, x\in W\cap V} \{v(x)-w(x)\} \geq u^*(y) - w(y) \geq \varphi(y) - w(y) \geq 0.$$

Since $s \geq \mu^V\psi$, it follows that

$$\liminf_{x\to y, x\in W\cap V} f(x) \geq 0 \qquad \text{for } y \in \partial W \cap \overline{V}.$$

If $y \in \partial V \cap W$, then

$$\liminf_{x\to y, x\in W\cap V} \{s(x)-w(x)\} \geq \psi(y) - w(y) \geq 0.$$

Since $\psi \leq u$ on ∂V, $\mu^V\psi \leq \mu^V u \leq v$ on V. Hence,

$$\liminf_{x\to y, x\in W\cap V} f(x) \geq 0 \qquad \text{for } y \in \partial V \cap W.$$

Since $V \cap W$ is an MP-set by Proposition 1.8, it follows that $f \geq 0$ on $V \cap W$. Taking the infimum of s and the supremum of w, we have

$$v + \mu^V\psi - \mu^W\varphi - \mu^V\psi \geq 0 \qquad \text{on } V \cap W,$$

i.e., $v \geq \mu^W\varphi$ on $V \cap W$. Hence, $u^* \geq \mu^W\varphi$ on $V \cap W$.

Finally, if $y \in \partial V \cap W$, then

$$u^*(y) = \min\{u(y), \liminf_{x\to y, x\in V\cap W} v(x)\} \geq (\mu^W\varphi)(y).$$

Thus, $u^* \geq \mu^W\varphi$ on W and the proposition is proved.

§2. Superharmonic functions and potentials

Throughout this section, let $(X, \mathcal{U})$ be a harmonic space and let $\mathcal{H} = \mathcal{H}_{\mathcal{U}}$. Given $U \in \mathcal{O}_X$, the set of all relatively compact resolutive set V such that $\overline{V} \subset U$ will be denoted by $\mathcal{O}_{rc}(U)$.

2-1. Superharmonic functions (cf. [11]

Let $U \in \mathcal{O}_X$. A function $u \in \mathcal{U}(U)$ is called superharmonic on U if for any $V \in \mathcal{O}_{rc}(U)$, $\mu^V u \in \mathcal{H}(U)$. Let $\mathcal{S}(U)$ be the set of all superharmonic functions on U. $\mathcal{S}(U)$ is a convex cone and closed under min. operation. If $-v \in \mathcal{S}(U)$, then v is said to be subharmonic on U.

Proposition 2.1. Every superharmonic function is finite on a dense set.

Proof. If $u \in \mathcal{S}(U)$ assumes $+\infty$ on an open set $W \neq \emptyset$, then choosing $V \in \mathcal{O}_{rc}(W)$ such that there exists $h \in \mathcal{H}(V) \cap \mathcal{C}(\overline{V})$ satisfying $h > 0$ on $\overline{V}$ (cf. Remark 1.1), we see that $u \geq nh$ on V for all n=1,2,.. and hence $\mu^V u \equiv +\infty$ on V, a contradiction.

Proposition 2.2 (a) A locally bounded hyperharmonic function is superharmonic; in particular $\mathcal{U}(U) \cap \mathcal{C}(U) \subset \mathcal{S}(U)$.

(b) If $u \in \mathcal{U}(U)$, $v \in \mathcal{S}(U)$ and $u \leq v$, then $u \in \mathcal{S}(U)$.

Proof. Let $u \in \mathcal{U}(U)$ and suppose it is locally bounded. For any $V \in \mathcal{O}_{rc}(U)$, there is $M > 0$ such that $|u| \leq M$ on $\overline{V}$. Then it follows from Axiom (BC) (or Proposition 1.4) that $\mu^V u \in \mathcal{H}(V)$. If $u \in \mathcal{U}(U)$, $v \in \mathcal{S}(U)$ and $u \leq v$, then for any $V \in \mathcal{O}_{rc}(U)$ and $\varphi \in \mathcal{C}(\partial V)$ with $\varphi \leq u$ on ∂V, $\mu^V \varphi \leq \mu^V v \in \mathcal{H}(V)$. Hence, again by Proposition 1.4., $\mu^V u \in \mathcal{H}(V)$.

Remark 2.1 If $(X, \mathcal{H})$ is a Brelot's harmonic space, then $u \in \mathcal{U}_{\mathcal{H}}(U)$ is superharmonic on U if and only if $u \not\equiv +\infty$ on each component of U; this fact is seen by Axiom 3.

Proposition 2.3. Let $u \in \mathcal{U}(U)$ and $V \in \mathcal{O}_{rc}(U)$. Put

$$u_V(y) = \begin{cases} u(y) & \text{if } y \in U \setminus \overline{V} \\ \mu^V u(y) & \text{if } y \in V \\ \min\{u(y), \liminf_{x \to y, x \in V} \mu^V u(x)\} & \text{if } y \in \partial V. \end{cases}$$

Then, $u_V \le u$ and for any $v \in -\mathcal{U}(U)$ such that $v \le u$ we have $v \le u_V$. If, furthermore, u has a subharmonic minorant on U, then $u_V \in \mathcal{U}(U)$. Thus, if $u \in \mathcal{S}(U)$ and has a subharmonic minorant on U, then $u_V \in \mathcal{S}(U)$.

Proof. Since $\mu^V u \le u$ on V, we have $u_V \le u$. Let $v \in -\mathcal{U}(U)$ and $v \le u$ on U. Then $v \le \mu^V u$ on V, so that $v \le u_V$ on U. Next, suppose $v \in -\mathcal{S}(U)$ and $v \le u$. Given $y \in \partial V$, we shall show that $u_V(y) > -\infty$. Then, by Proposition 1.9, we conclude that $u_V \in \mathcal{U}(U)$.
Since v is locally bounded above, we can choose an MP-set V' containing y such that there exists $h \in \mathcal{H}(V')$ satisfying $v \le h$ on V'. Choose $W \in \mathcal{O}_{rc}(V')$ such that $y \in W$, and consider the function

$$v_W(x) = \begin{cases} v(x) & \text{if } x \in U \setminus W \\ \mu^W v(x) & \text{if } x \in W \\ \max\{v(x), \limsup_{y \to x, y \in W} \mu^W v(y)\} & \text{if } x \in \partial W. \end{cases}$$

Since $\mu^W v \le h$ on W, $v_W \le h$ on W', and hence $v_W(x) < +\infty$ for all $x \in \partial W$. Then, by Proposition 1.9., $v_W \in -\mathcal{U}(W)$. Now, $v \le u$ implies $v_W \le u$, and hence $v_W \le u_V$. Since $v \in -\mathcal{S}(U)$, $v_W(y) > -\infty$. Hence $u_V(y) > -\infty$.

The last assertion of the proposition now follows from Proposition 2.2.(b).

Let $U \in \mathcal{O}_X$. A non-empty family $\mathcal{V} \subset \mathcal{U}(U)$ is called a Perron set on U if it satisfies the following two conditions:

(a) $\mathcal{V}$ is lower directed and possesses a subharmonic minorant;

(b) for any $x \in U$, there is $V \in \mathcal{O}_{rc}(U)$ containing x such that $u_V \in \mathcal{V}$ for all $u \in \mathcal{V}$ and $\mu^V u_o \in \mathcal{H}(V)$ for some $u_o \in \mathcal{V}$.

Theorem 2.1. (Perron) If $\mathcal{V}$ is a Perron set on U, then $\inf \mathcal{V} \in \mathcal{H}(U)$.

Proof. Let V be a set satisfying the condition given in (b) for Perron set. Then

$$\mathcal{V}_V = \{u_V \mid u \in \mathcal{V}\}$$

is also lower directed and $\inf \mathcal{V}_V = \inf \mathcal{V}$. By Proposition 1.4, we see that $\inf \mathcal{V}_V \in \mathcal{H}(V)$. Since such V's cover U, $\inf \mathcal{V} \in \mathcal{H}(U)$.

2-2. Potentials (cf. [11])

Let $U \in \mathcal{O}_X$. $p \in \mathcal{S}(U)$ is called a potential on U if $p \geq 0$ on U and if $h \in \mathcal{H}(U)$, $h \leq p$ imply $h \leq 0$. The set of all potentials on U will be denoted by $\mathcal{P}(U)$.

If $u \in \mathcal{U}(U)$, $u \geq 0$ and $u \leq p$ for some $p \in \mathcal{P}(U)$, then $u \in \mathcal{P}(U)$. It follows that $\mathcal{P}(U)$ is closed under min. operation.

Theorem 2.2. (F. Riesz' decomposition theorem) If $u \in \mathcal{S}(U)$ has a subharmonic minorant on U, then u is uniquely decomposed as $u = h + p$ with $h \in \mathcal{H}(U)$ and $p \in \mathcal{P}(U)$. Furthermore,

$$h = \sup \{v \in -\mathcal{U}(U) \mid v \leq u\}, \tag{2.1}$$

in particular, h is the greatest harmonic minorant of u.

Proof. Let

$$\mathcal{V} = \{u_{V_1 V_2 \dots V_n} \mid V_j \in \mathcal{O}_{rc}(U),\ j = 1,\dots,n;\ n = 1,2,\dots\}.$$

Then, it is easy to see that $\mathcal{V}$ is a Perron set. Hence, by Theorem 2.1., $h = \inf \mathcal{V} \in \mathcal{H}(U)$. Obviously, $h \leq u$. By Proposition 2.3, we see that (2.1) holds. Put $p = u-h$. Then $p \in \mathcal{S}(U)$ and $p \geq 0$ on U. If $h_1 \in \mathcal{H}(U)$ and $h_1 \leq p$, then $h + h_1 \leq u$. Hence by (2.1), $h + h_1 \leq h$, i.e., $h_1 \leq 0$. Hence $p \in \mathcal{P}(U)$. If $u = \tilde{h} + \tilde{p}$ is another decompostion with $\tilde{h} \in \mathcal{H}(U)$ and $\tilde{p} \in \mathcal{P}(U)$, then $\tilde{h} - h \leq p$ and $h - \tilde{h} \leq \tilde{p}$ imply $h = \tilde{h}$. Thus, the decomposition is unique.

Corollary 2.1. If $p \in \mathcal{P}(U)$, $v \in -\mathcal{U}(U)$ and $v \leq p$, then $v \leq 0$.

Corollary 2.2. $\mathcal{P}(U)$ is a convex cone.

Proof. If $p \in \mathcal{P}(U)$ and $\alpha > 0$, then clearly $\alpha p \in \mathcal{P}(U)$. If $p_1, p_2 \in \mathcal{P}(U)$, then $p_1+p_2 \in \mathcal{S}(U)$ and $p_1+p_2 \geq 0$. If $h \in \mathcal{H}(U)$ and $h \leq p_1+p_2$, then $h-p_1 \in -\mathcal{S}(U)$ and $h-p_1 \leq p_2$, so that $h-p_1 \leq 0$ by the above corollary. Hence $h \leq 0$, and thus $p_1+p_2 \in \mathcal{P}(U)$.

Proposition 2.4. Let $p_n \in \mathcal{P}(U)$, $n = 1,2,\ldots$, and suppose

$$\Sigma_{n=1}^{\infty} p_n \in \mathcal{S}(U). \text{ Then } \Sigma_{n=1}^{\infty} p_n \in \mathcal{P}(U).$$

Proof. Put $p = \Sigma_{n=1}^{\infty} p_n$ and let $h \in \mathcal{H}(U)$ and $h \leq p$. For any k, $h - \Sigma_{n=k+1}^{\infty} p_n$ is subharmonic and dominated by $\Sigma_{n=1}^{k} p_n \in \mathcal{P}(U)$. Hence, $h \leq \Sigma_{n=k+1}^{\infty} p_n$. Since p is superharmonic by assumption, it is finite on a dense subset of U (Proposition 2.1). It follows that $h \leq 0$ on a dense subset of U. Since h is continuous, $h \leq 0$. Thus $p \in \mathcal{P}(U)$.

Proposition 2.5. Let $p \in \mathcal{P}(U)$ and suppose p is harmonic on $U \setminus K$ for some compact set K in U. Let $u \in \mathcal{U}(U)$ and suppose $u \geq 0$ on U and

$$\lim_{x \to \xi,\, x \in U\setminus K} \{u(x) - p(x)\} \geq 0$$

for all $\xi \in \partial K$. Then $u \geq p$ on $U \setminus K$.

Proof. Consider the function

$$v = \begin{cases} 0 & \text{on } K \\ \min(0, u-p) & \text{on } U\setminus K. \end{cases}$$

By the assumption of the proposition, we see that v is lower semicontinuous on U. Hence, by Proposition 1.7, $v \in \mathcal{U}(U)$. Obviously $v \geq -p$. Hence $v \geq 0$ on U, i.e., $u \geq p$ on $U \setminus K$.

2-3. Reduced functions (cf. [11]).

Let $U \in \mathcal{O}_X$. Given a function f on U, we define

$$R_U f = \inf \{u \in \mathcal{U}(U) \mid u \geq f \text{ on } U\}.$$

Obviously, $R_U f \geq f$, $f \leq g$ implies $R_U f \leq R_U g$, $R_U(\alpha f) = \alpha R_U f$ for $\alpha \geq 0$, $R_U(f+g) \leq R_U f + R_U g$. If $f \in \mathcal{U}(U)$, then $R_U f = f$.

Proposition 2.6. (i) If f is lower semicontinuous $]-\infty,+\infty]$-valued on U, then $R_U f \in \mathcal{U}(U)$.

(ii) If f is lower semicontinuous $]-\infty,+\infty]$-valued and has a superharmonic majorant on U, then $R_U f \in \mathcal{S}(U)$, $R_U f$ is continuous at any point where f is continuous, and $R_U f$ is harmonic on any open set where f is either subharmonic or continuous and strictly smaller than $R_U f$.

Proof. Fix U and write Rf for $R_U f$.

(i) Let $u(x) = \liminf_{y\to x} Rf(y)$ for every $x \in U$. Then u is lower semicontinuous and $u \geq f$, in particular u is $]-\infty,+\infty]$-valued. Let $V \in \mathcal{O}_{rc}(U)$. For any $v \in \mathcal{U}(U)$ satisfying $v \geq f$ on U, $v \geq u$ on U, so that $\mu^V u \leq \mu^V v \leq v$ on V. Hence $\mu^V u \leq Rf$ on V. Since $\mu^V u$ is lower semicontinuous on V, it follows that $\mu^V u \leq u$ on V. Hence $u \in \mathcal{U}(U)$. On the other hand, $Rf \geq f$ implies $u \geq f$. Hence $Rf \leq u \leq v$. It follows that $Rf = u \in \mathcal{U}(U)$.

(ii) Let v_o be a superharmonic majorant of f. Then $Rf \leq v_o$, and hence $Rf \in \mathcal{S}(U)$ by Proposition 2, (b).

Suppose f is continuous at $x_o \in U$. Choose $V' \in \mathcal{O}_X$ such that $x_o \in V'$ and there is $h \in \mathcal{H}(V')$ with $h(x_o) = 1$. Given $\varepsilon > 0$, we can choose $V \in \mathcal{O}_{rc}(V' \cap U)$ such that $x_o \in V$,

$$f \leq (f(x_o) + \varepsilon)h \qquad \text{on } \overline{V}$$

and

$$Rf \geq (f(x_o) - \varepsilon)h \qquad \text{on } \overline{V}.$$

Then, $\mu^V(Rf) \geq (f(x_o)-\varepsilon)h$ on V, so that $(Rf)_V \geq (f(x_o)-\varepsilon)h$ on $\overline{V}$.

Hence

$$f \leq (f(x_o) + \varepsilon)h \leq (Rf)_V + 2\varepsilon h \qquad \text{on } \overline{V}.$$

Let

$$u^*(y) = \begin{cases} Rf(y) & \text{if } y \in U \setminus \overline{V} \\ \min(Rf(y),\ (Rf)_V(y) + 2\varepsilon h(y)) & \text{if } y \in V \\ \min\{Rf(y),\ \liminf\limits_{x \to y, x \in V} [\mu^V(Rf)(x) + 2\varepsilon h(x)]\} & \text{if } y \in \partial V. \end{cases}$$

Then, $f \leq u^*$ and hence by Proposition 1.9, $u^* \in \mathcal{U}(U)$. Hence $u^* \geq Rf$, so that

$$(Rf)_V + 2\varepsilon h \geq Rf \qquad \text{on } V.$$

Since Rf is superharmonic, $(Rf)_V(x_o) < +\infty$, so that $Rf(x_o) < +\infty$. Furthermore, since $(Rf)_V$ is continuous at x_o, we have

$$\limsup_{y \to x_o} Rf(y) \leq (Rf)_V(x_o) + 2\varepsilon \leq Rf(x_o) + 2\varepsilon.$$

Since ε is arbitrary and Rf is lower semicontinuous, it follows that Rf is continuous at x_o.

Next, suppose f is subharmonic on $W \subset U$ ($W \in \mathcal{O}_X$). For each $x \in W$, we can choose $V \in \mathcal{O}_{rc}(W)$ containing x. Since $f \leq Rf$, $f \leq (Rf)_V$ by Proposition 2.3. Since $(Rf)_V \in \mathcal{U}(U)$, it follows that $Rf = (Rf)_V$. Therefore, Rf is harmonic on V. Since $x \in W$ is arbitrary, Rf is harmonic on W.

Finally, suppose f is continuous and $f < Rf$ on $W \subset U$ ($W \in \mathcal{O}_X$). For each $x \in W$, choose $V' \in \mathcal{O}_X$ such that $x \in V'$ and there is $h \in \mathcal{H}(V')$ with $h(x) = 1$. Since f is continuous and Rf is lower semi-continuous, we can find $V \in \mathcal{O}_{rc}(V' \cap W)$ such that $x \in V$ and $f < \alpha h < Rf$ on $\overline{V}$ for some $\alpha \in \mathbb{R}$. Then $f \leq (Rf)_V$, and it follows that $Rf = (Rf)_V$ as above. Thus Rf is harmonic on V, and hence on W.

2-4. P-sets (cf. [11])

A non-empty open set U in X is called a P-set if for any $x \in U$, there is $p \in \mathcal{P}(U)$ such that $p(x) > 0$.

Proposition 2.7. If U is a P-set and if f is a non-negative bounded lower semicontinuous function with compact support on U, then $R_U f \in \mathcal{P}(U)$.

Proof. Let $K = \mathrm{Supp}\, f$. For each $x \in K$, there is $p_x \in \mathcal{P}(U)$ such that $p_x(x) > 0$. $V_x = \{y \in U \mid p_x(y) > 0\}$ is an open subset of U. Since K is compact, we can find a finite number of points $x_1, \ldots, x_n \in K$ such that $V_{x_1} \cup \ldots \cup V_{x_n} \supset K$. Then $p = p_{x_1} + \ldots + p_{x_n}$ is a potential on U and $\inf_K p > 0$. Hence $\alpha p \geq f$ for some constant $\alpha > 0$. Then $0 \leq R_U f \leq \alpha p$, which implies that $R_U f \in \mathcal{P}(U)$ in view of Proposition 2.6.

Proposition 2.8. If U is a P-set, then there exists $p \in \mathcal{P}(U)$ which is continuous and strictly positive everywhere on U.

Proof. Let $\{W_n\}$ be an exhaustion of U and for each n choose $f_n \in \mathbf{C}(U)$ such that $f_n \geq 0$ on U, $f_n = 1$ on W_n and $\mathrm{Supp} f_n \subset W_{n+1}$. By Propositions 2.6 and 2.7, $p_n = R_U f_n$ is a continuous potential on U and $p_n \geq f_n$ for each n. Put $\alpha_n = \sup_{\overline{W}_n} p_n$. Then $0 < \alpha_n < +\infty$. Hence

$$p = \sum_{n=1}^{\infty} \frac{1}{2^n \alpha_n} p_n$$

is a continuous hyperharmonic function, so that $p \in \mathcal{P}(U)$ by Propositions 2.2 and 2.4. Obviously, $p(x) > 0$ for all $x \in U$.

Proposition 2.9. Let U be a P-set. Then for any $s \in \mathcal{S}^+(U)$, there exists an increasing sequence $\{p_n\}$ of continuous potentials such that each p_n is harmonic outside a compact set in U and $s = \lim_{n\to\infty} p_n$.

Proof. There exists an increasing sequence $\{f_n\}$ of non-negative continuous functions on U such that each f_n has compact support in U and $f_n \uparrow s$ as $n \to \infty$. Put $p_n = R_U f_n$. Then, by Propositions 2.6 and 2.7, each p_n is a continuous potential on U and harmonic outside supp f_n. Since $f_n \le p_n \le s$, $s = \lim_{n\to\infty} p_n$.

Proposition 2.10. If U is a P-set, then $\mathcal{P}(U) \cap \mathcal{C}(U)$ separates points of U.

Proof. Let $x, y \in U$ and $x \neq y$. Let $p \in \mathcal{P}(U)$ be continuous and strictly positive on U, whose existence is assured by Proposition 2.8. Consider the family

$$\mathcal{V} = \{ p_{V_1 \ldots V_n} \mid V_j \in \mathcal{O}_{rc}(U), \text{ either } x \notin \overline{V}_j \text{ or } y \notin \overline{V}_j \text{ for each } j \}.$$

Then, $\mathcal{V}$ is a Perron set, and hence $\inf \mathcal{V} \in \mathcal{H}(U)$ by Theorem 2.1. Since $0 \le \inf \mathcal{V} \le p$, it follows that $\inf \mathcal{V} = 0$. Hence, we can choose $V_j \in \mathcal{O}_{rc}(U)$, $j = 1,2,\ldots$, such that either $x \notin \overline{V}_j$ or $y \notin \overline{V}_j$ for each j and, putting $p_n = p_{V_1 \ldots V_n}$, $p_n(x) \to 0$ and $p_n(y) \to 0$ $(n \to \infty)$. Let k be the smallest number such that either $p_k(x) < p(x)$ or $p_k(y) < p(y)$ occurs. Then either

$$p_k(x) < p(x) \quad \text{and} \quad p_k(y) = p(y)$$

or

$$p_k(x) = p(x) \quad \text{and} \quad p_k(y) < p(y).$$

Therefore, p and p_k separate points x and y. By using Proposition 2.9, we can find $q \in \mathcal{P}(U) \cap \mathcal{C}(U)$ such that p and q separate x and y.

Proposition 2.11. Let U be a P-set and $V \in \mathcal{O}_{rc}(U)$. Then, for each $x \in V$, there exists $p \in \mathcal{P}(U) \cap \mathcal{C}(U)$ such that

$$\mu^V p(x) < p(x).$$

Proof. If $\mu^V_x = 0$, then we may take any $p \in \mathcal{P}(U) \cap \mathcal{C}(U)$ such that $p(x) > 0$. Suppose $\mu^V_x \neq 0$. Let $y \in \operatorname{Supp} \mu^V_x$. Then $y \neq x$. Hence, by the previous proposition, there are $q_1, q_2 \in \mathcal{P}(U) \cap \mathcal{C}(U)$ such that

$$q_1(x) q_2(y) < q_1(y) q_2(x).$$

Put

$$p = \min(q_2(x) q_1,\ q_1(x) q_2).$$

Then, $p \in \mathcal{P}(U) \cap \mathcal{C}(U)$. Since $p(y) < q_1(y) q_2(x)$ and $y \in \operatorname{Supp} \mu^V_x$,

$$(\mu^V p)(x) < (\mu^V q_1)(x)\ q_2(x) \leq q_1(x) q_2(x) = p(x).$$

Proposition 2.12. Any non-empty open subset of a P-set is a P-set.

Proof. Let U be a P-set and U' be a non-empty open subset of U. Let $x \in U'$ and choose $V \in \mathcal{O}_{rc}(U')$ containing x. By the previous proposition there is $p \in \mathcal{P}(U)$ such that $\mu^V p(x) < p(x)$. By Theorem 2.2,

$$p|U' = h + p' \qquad \text{with } h \in \mathcal{H}(U') \text{ and } p' \in \mathcal{P}(U').$$

Since $\mu^V h = h$, we have $\mu^V p'(x) < p'(x)$, so that $p'(x) > 0$. Hence U' is a P-set.

Proposition 2.13. Let U be a relatively compact non-empty open set such that $\overline{U}$ is contained in an MP-set. If there exists $h \in \mathcal{H}(U) \cap \mathcal{C}(\overline{U})$ such that $h > 0$ on U, then U is a P-set.

Proof. Suppose U is not a P-set. Then there exists $x_o \in U$ such that every potential on U vanishes at x_o. Let

$$\mathcal{F} = \{h \in \mathcal{H}(U) \mid h \geq 0, \text{ bounded}, h(x_o) = 1\}.$$

By assumption $\mathcal{F} \neq \emptyset$. If $h_1, h_2 \in \mathcal{F}$, then $\min(h_1, h_2) \in \mathcal{S}(U)$, so that

$$\min(h_1, h_2) = h' + p \qquad \text{with } h' \in \mathcal{H}(U),\ p \in \mathcal{P}(U).$$

Then, $h' \in \mathcal{F}$, since $p(x_o) = 0$. Obviously, $h' \le \min(h_1, h_2)$. Thus $\mathcal{F}$ is lower directed. Hence $h_o = \inf \mathcal{F}$ belongs to $\mathcal{H}(U)$ by Proposition 1.4. Clearly, $h_o \in \mathcal{F}$, i.e., h_o is the smallest function in $\mathcal{F}$. Now, let U' be an MP-set containing $\overline{U}$ and put

$$v(y) = \begin{cases} h_o(y) & \text{if } y \in U \\ 0 & \text{if } y \in U' \setminus \overline{U} \\ \limsup\limits_{z \to y, z \in U} h_o(z) & \text{if } y \in \partial U. \end{cases}$$

Then, v is upper semicontinuous on U', $v \ge 0$ and bounded. Let $V \in \mathcal{O}_{rc}(U')$ and suppose $x_o \notin \overline{V}$. For any $s \in \overline{\mathcal{U}}_v^V$, the function

$$w = \begin{cases} h_o & \text{on } U \setminus V \\ \inf(s, h_o) & \text{on } U \cap V \end{cases}$$

belongs to $\mathcal{S}(U)$ by Proposition 1.7. Let $w = h_1 + p_1$ with $h_1 \in \mathcal{H}(U)$ and $p_1 \in \mathcal{P}(U)$. Since $w \ge 0$, we have $h_1 \ge 0$. Since $w \le h_o$, we have $h_1 \le h_o$, so that h_1 is bounded. Since $p_1(x_o) = 0$, $h_1(x_o) = h_o(x_o) = 1$. Hence $h_1 \in \mathcal{F}$, which implies $h_1 = h_o$. Therefore, $s \ge h_o$ on $U \cap V$, and hence $\mu^V v \ge h_o$ on $U \cap V$. It follows that $\mu^V v \ge v$ on V, so that v is subharmonic on U'. Since U' is an MP-set, $v \le 0$ on U, i.e., $h_o = 0$, contradicting $h_o(x_o) = 1$.

Corollary 2.3. Any point in X has an open neighborhood which is a P-set.

Remark 2.2. From this Corollary and Proposition 2.10, it follows that any harmonic space satisfies Axiom (S) for Bauer's harmonic space (cf. 1-3). Hence, a harmonic space having a base consisting of regular sets is Bauer's harmonic space; in particular, a Brelot's harmonic space is a Bauer's harmonic space.

Proposition 2.14. If U is a P-set, then the following minimum principle holds: If $u \in \mathcal{U}(U)$ and $u \ge 0$ outside a compact set in U, then $u \ge 0$ on U.

<u>Proof.</u> Let K be a compact set in U such that $u \geq 0$ on $U \setminus K$. For each $x \in K$, choose $V_x \in \mathcal{O}_{rc}(U)$ which contains x. By Proposition 2.11 , there exists $p_x \in \mathcal{P}(U) \cap \mathcal{C}(U)$ such that $\mu^{V_x} p_x(x) < p_x(x)$. Since both $\mu^{V_x} p_x$ and p_x are continuous at x, there is an open neighborhood W_x of x such that $W_x \subset V_x$ and $\mu^{V_x} p_x < p_x$ on W_x. Choose $x_1, \ldots, x_n \in K$ such that $W_{x_1} \cup \ldots \cup W_{x_n} \supset K$ and put $p = \Sigma_{i=1}^{n} p_{x_i}$.

Obviously, $p(x) > 0$ for all $x \in K$.

Now suppose $u(x_o) < 0$ for some $x_o \in K$. Then $\alpha \equiv \sup_K (-u/p) > 0$. Since $-u/p$ is upper semicontinuous on K, there is $z \in K$ such that $\alpha p(z) + u(z) = 0$. Let $z \in W_{x_j}$ and $V = V_{x_j}$. Since $\alpha p + u \geq 0$ on U,

$$0 = \alpha p(z) + u(z) > \alpha \mu^V p(z) + \mu^V u(z) = \mu^V (\alpha p + u)(z) \geq 0,$$

which is a contradiction. Hence $u \geq 0$ on U.

<u>Corollary 2.4.</u> Let U be a P-set, $p \in \mathcal{P}(U)$, $u \in \mathcal{S}(U)$ and $u \geq 0$. If $u \leq p$ outside a compact set in U, then $u \in \mathcal{P}(U)$.

<u>Proof.</u> Let $h \in \mathcal{H}(U)$ and $h \leq u$. Then $p - h \geq 0$ outside a compact set in U. Hence by the above proposition $p - h \geq 0$ on U, which implies $h \leq 0$.

<u>Proposition 2.15.</u> An open set U such that $\overline{U}$ is contained in a P-set is an MP-set.

<u>Proof.</u> Let U' be a P-set such that $\overline{U} \subset U'$. Let $u \in \mathcal{U}(U)$ satisfy $u \geq 0$ on $U \setminus K$ for some compact set K in X and

$$\liminf_{x \to \xi, x \in U} u(x) \geq 0 \qquad \text{for all } \xi \in \partial U.$$

Put

$$u^* = \begin{cases} \min(u,o) & \text{on } U \\ 0 & \text{on } U' \setminus U. \end{cases}$$

Then $u^* \in \mathcal{U}(U')$ by Proposition 1.7 and $u^* = 0$ on $U' \setminus (\bar{U} \cap K)$. Since $\bar{U} \cap K$ is compact in U', Proposition 2.14 implies that $u^* \geq 0$ on U'. Hence $u \geq 0$ on U. Therefore U is an MP-set.

Theorem 2.3. Let U be a P-set. Then, for any $f \in \mathcal{C}_0^+(U)$ and $\varepsilon > 0$, there exist $p, q \in \mathcal{P}(U) \cap \mathcal{C}(U)$ satisfying the following two conditions:

(a) p, q are harmonic on $U \setminus \operatorname{Supp} f$.

(b) $0 \leq p - q \leq f \leq p - q + \varepsilon$ on U; in particular $p - q$ has compact support.

Proof. Let

$$A = \{x \in U \mid f(x) \geq \tfrac{\varepsilon}{3}\}.$$

A is a compact set in U. We consider the family

$$\mathcal{F} = \{g \in \mathcal{C}(A) \mid g = p - q \text{ on } A \text{ with } p, q \in \mathcal{P}(U) \cap \mathcal{C}(U)\}.$$

Then, $\mathcal{F}$ is a linear subspace of $\mathcal{C}(A)$. If $g = p - q$ on A with $p, q \in \mathcal{P}(U) \cap \mathcal{C}(U)$, then $\min(g, 0) = \min(p, q) - q$ on A and $\min(p, q) \in \mathcal{P}(U) \cap \mathcal{C}(U)$. Hence $\min(g, 0) \in \mathcal{F}$. It follows that $\mathcal{F}$ is closed under max. and min. operations. By Proposition 2.10, we see that $\mathcal{F}$ separates points of A. Hence Stone's theorem implies that $\mathcal{F}$ is dense in $\mathcal{C}(A)$. Therefore, we find $g \in \mathcal{F}$ such that

$$|f - \tfrac{2}{3}\varepsilon - g| < \tfrac{\varepsilon}{3} \quad \text{on } A.$$

Put $g = p_0 - q_0$ on A with $p_0, q_0 \in \mathcal{P}(U) \cap \mathcal{C}(U)$. Consider the functions

$$p_1 = \begin{cases} p_0 & \text{on } A \\ \min(p_0, q_0) & \text{on } U \setminus A \end{cases} \quad \text{and} \quad q_1 = \min(p_1, q_0).$$

Since $f = \varepsilon/3$ on ∂A, $g \leq 0$ on ∂A, that is $p_0 \leq q_0$ on ∂A. Since p_0, q_0 are continuous, it follows that p_1 is continuous. Hence $p_1 \in \mathcal{P}(U) \cap \mathcal{C}(U)$ by Proposition 1.7 and the fact that $0 \leq p_1 \leq p_0$. Then, we also have $q_1 \in \mathcal{P}(U) \cap \mathcal{C}(U)$. On $U \setminus A$, $p_1 = q_1$. On A, we have $p_1 - q_1 = p_0 - \min(p_0, q_0) = \max(g, 0)$.

Since $f-\varepsilon < g < f$ on A, it follows that

(2.2) $$0 \leq p_1 - q_1 \leq f \leq p_1 - q_1 + \varepsilon \quad \text{on } A.$$

Choose $\varphi \in \mathcal{C}_0(U)$ such that $0 \leq \varphi \leq 1$ on U, $\varphi = 1$ on A and Supp $\varphi \subset$ Supp f. Put $p = R_U(\varphi p_1)$ and $q = R_U(\varphi\, q_1)$. Then $p, q \in \mathcal{P}(U) \cap \mathcal{C}(U)$ and are harmonic on $U \setminus$ Supp f by Propositions 2.6 and 2.7, $p \geq q$, $\varphi p_1 \leq p \leq p_1$ and $\varphi\, q_1 \leq q \leq q_1$. Since $p = p_1 = q_1 = q$ on ∂A and $p_1 = q_1$ on $U \setminus A$,

$$p' = \begin{cases} p & \text{on } A \\ q & \text{on } U\setminus A \end{cases}$$

is a potential by Proposition 1.7 and $p' \geq \varphi p_1$. Hence $p' \geq p$, which implies that $p = q$ on $U \setminus A$. On the other hand, since $p = p_1$ and $q = q_1$ on A, (2.2) shows that

$$0 \leq p - q < f < p - q + \varepsilon \quad \text{on } A.$$

Therefore, the above p and q are the required functions.

2-5. The space $\mathcal{R}(U)$ (cf. [26]; also [14] and [32])

Hereafter we denote the space $\mathcal{S}(U) \cap \mathcal{C}(U)$ by $\mathcal{S}_C(U)$ for $U \in \mathcal{O}_X$.

For $U \in \mathcal{O}_X$, we define

$$\mathcal{R}(U) = \left\{ f \in \mathcal{C}(U) \;\middle|\; \begin{array}{l} \text{for any } x \in U, \text{ there is } V_x \in \mathcal{O}_X \text{ such that} \\ x \in V_x \subset U \text{ and } f|V_x \in \mathcal{S}_C(V_x) - \mathcal{S}_C(V_x) \end{array} \right\}.$$

It is easy to see that $\mathcal{R}(U)$ is a linear subspace of $\mathcal{C}(U)$ and $\mathcal{R}: U \mapsto \mathcal{R}(U)$ is a sheaf of functions on X. Obviously,

$$\mathcal{H}(U) \subset \mathcal{S}_C(U) - \mathcal{S}_C(U) \subset \mathcal{R}(U).$$

Proposition 2.16. $\mathcal{R}(U)$ is closed under max. and min. operations.

Proof. Let $f \in \mathcal{R}(U)$. For each $x \in U$, choose $V \in \mathcal{O}_x$ such that $x \in V \subset U$ and $u, v \in \mathcal{S}_c(V)$ such that $f|V = u - v$. Then

$$\max(f,0)|V = u - \min(u,v)$$

and $\min(u,v) \in \mathcal{S}_c(V)$.

Proposition 2.17. If K is a compact set in U, then there exists $f \in \mathcal{R}(U)$ such that $f \geq 0$ on U, $f > 0$ on K and Supp f is compact and contained in U. If, in particular, $1 \in \mathcal{R}(U)$, then we can choose f so that $f = 1$ on K.

Proof. For each $x \in K$, we can choose a P-set V_x such that $x \in V_x \subset U$ (Corollary 2.3). Let W_x be a relatively compact open neighborhood of x such that $\overline{W}_x \subset V_x$. Choose $f_x \in \mathcal{C}_o^+(V_x)$ such that $f_x = 1$ on $\overline{W}_x$. Then, by Theorem 2.3, there exist $p_x, q_x \in \mathcal{P}(V_x) \cap \mathcal{C}(V_x)$ such that $p_x - q_x \geq 0$ on V_x, $p_x - q_x \geq \frac{1}{2}$ on W_x and $\mathrm{Supp}(p_x - q_x)$ is compact in V_x. Put

$$g_x = \begin{cases} p_x - q_x & \text{on } V_x \\ 0 & \text{on } U \backslash V_x. \end{cases}$$

Then, $g_x \in \mathcal{R}(U)$. Choose $x_1, \ldots, x_n \in K$ such that $W_{x_1} \cup \ldots \cup W_{x_n} \supset K$ and put $f = g_{x_1} + \ldots + g_{x_n}$. Then $f \geq 0$ on U, $f > 0$ on K and Supp f is compact in U.

In the case where $1 \in \mathcal{R}(U)$, $f_1 = \min(\frac{f}{\alpha}, 1)$ belongs to $\mathcal{R}(U)$ by Proposition 2.16, and $f_1 = 1$ on K, where $\alpha = \inf_K f > 0$.

Corollary 2.5. There exists $f \in \mathcal{R}(X)$ which is strictly positive everywhere on X.

Proof. Let $\{U_n\}$ be an exhaustion of X and choose $f_n \in \mathcal{R}(X)$ such that $f_n \geq 0$ on X, $f_n > 0$ on $\overline{U}_n \backslash U_{n-1}$ and Supp $f_n \subset U_{n+1} \backslash \overline{U}_{n-2}$, $n = 1, 2, \ldots$, where $U_o = U_{-1} = \emptyset$. Then $f = \Sigma_{n=1}^{\infty} f_n$ is the required function.

Corollary 2.6. $\mathcal{R}(U)$ is dense in $\mathcal{C}(U)$ with respect to the locally uniform convergence topology.

Proof. By Proposition 1.7 , we see that $\mathcal{R}(U)$ separates points of U. Since $\mathcal{R}(U)$ is a vector lattice with respect to the max. and min. operations, Stone's theorem implies this corollary.

Lemma 2.1. Let $u \in \mathcal{S}_C(U)$, $v \in -\mathcal{S}_C(U)$, $u > 0$ on U and $v \geq 0$ on U. Then $v^n u^{1-n} \in -\mathcal{S}_C(U)$ for $n \geq 1$.

Proof. Obviously, $v^n u^{1-n} \in \mathcal{C}(U)$. Let $V \in \mathcal{O}_{rc}(U)$. For any $x \in V$, by Hölder's inequality we have

$$v^n(x) \leq (\int v \, d\mu_x^V)^n \leq \{\int \frac{v^n}{u^{n-1}} \, d\mu_x^V\}\{\int u d\mu_x^V\}^{n-1}$$

$$\leq u^{n-1}(x) \int \frac{v^n}{u^{n-1}} \, d\mu_x^V ,$$

i.e., $v^n u^{1-n} \leq \mu^V(v^n u^{1-n})$ on V. Hence $v^n u^{1-n} \in -\mathcal{S}_C(U)$.

Lemma 2.2. Let $u \in \mathcal{S}_C(U)$, $v \in -\mathcal{S}_C(U)$ and $0 \leq v < u$ on U. Then $u^2(u+v)^{-1} \in \mathcal{S}_C(U) - \mathcal{S}_C(U)$.

Proof. We have

$$u^2(u+v)^{-1} = u(1 + \frac{v}{u})^{-1} = u \sum_{n=o}^{\infty} (-1)^n (\frac{v}{u})^n = \sum_{m=o}^{\infty} \frac{v^{2m}}{u^{2m-1}} - \sum_{m=o}^{\infty} \frac{v^{2m+1}}{u^{2m}},$$

where each series converges locally uniformly on U, since $0 \leq v/u < 1$ on U. Put

$$w_1 = - \sum_{m=o}^{\infty} \frac{v^{2m}}{u^{2m-1}} \quad \text{and} \quad w_2 = - \sum_{m=o}^{\infty} \frac{v^{2m+1}}{u^{2m}} .$$

Then, $w_1, w_2 \in \mathcal{C}(U)$. By Lemma 2.1 , partial sums of the series defining w_1, w_2 are superharmonic. Since the convergence is monotone and locally uniform, it follows that $w_1, w_2 \in \mathcal{S}_C(U)$.

<u>Lemma 2.3.</u> Let $u \in \mathcal{S}_C(U)$ and $u > 0$ on U. Then, for any $f,g \in \mathcal{R}(U)$, $fgu^{-1} \in \mathcal{R}(U)$.

<u>Proof.</u> For each $x \in U$ choose an open neighborhood V of x such that there is $h \in \mathcal{H}(V)$ with $\inf_V h \geq 1$, $f|V = s_1 - s_2$ and $g|V = t_1 - t_2$ with $s_1, s_2, t_1, t_2 \in \mathcal{S}_C(V) \cap \mathcal{C}(\overline{V})$. Let

$$\alpha = \max(\sup_U s_1, \sup_U s_2), \qquad \beta = \max(\sup_U t_1, \sup_U t_2)$$

$$v_i = \alpha h - s_i \quad , \quad w_i = \beta h - t_i \quad (i = 1,2).$$

Then, $f|V = v_2 - v_1$ and $g|V = w_2 - w_1$, so that

$$(2.3) \qquad \frac{fg}{u} = \frac{v_1 w_1 + v_2 w_2}{u} - \frac{v_1 w_2 + v_2 w_1}{u} \quad \text{on } V.$$

Since $v_i, w_i \in -\mathcal{S}_C(V)$ and $v_i, w_i \geq 0$ on V $(i = 1,2)$, by Lemma 2.1., we have $v_i^2 u^{-1}$, $w_j^2 u^{-1}$, $(v_i + w_j)^2 u^{-1} \in -\mathcal{S}_C(V)$ for $i,j = 1,2$. Hence, $v_i w_j u^{-1} \in \mathcal{S}_C(V) - \mathcal{S}_C(V)$ $(i,j = 1,2)$. Therefore (2.3) shows that $fgu^{-1} \in \mathcal{S}_C(V) - \mathcal{S}_C(V)$. Since x is arbitrary, it follows that $fgu^{-1} \in \mathcal{R}(U)$.

<u>Proposition 2.18.</u> (Hansen) If $f,g,h \in \mathcal{R}(U)$ and $h > 0$ on U, then $fgh^{-1} \in \mathcal{R}(U)$.

<u>Proof.</u> For each $x \in U$, choose an open neighborhood V of x such that there is $h_o \in \mathcal{H}(V)$ with $h_o(x) = 1$ and $h|V = u_1 - u_2$ with $u_i \in \mathcal{S}_C(V)$, $i = 1,2$. Let $\lambda = \frac{1}{3} u_1(x) + \frac{2}{3} u_2(x)$ and consider the functions

$$v = \lambda h_o - u_2 \quad \text{and} \quad u = u_1 - \lambda h_o .$$

Then, $v \in -\mathcal{S}_C(V)$ and $u \in \mathcal{S}_C(V)$. Since $v(x) = \frac{1}{3} h(x) > 0$ and $u(x) = \frac{2}{3} h(x)$, there is another open neighborhood W of x such that $W \subset V$ and $0 < v < u$ on W. By Lemma 2.2 , $u^2(u+v)^{-1} \in \mathcal{R}(W)$.

Since $u+v = u_1 - u_2 = h$,

$$fgh^{-1} = fg(u+v)^{-1} = (fgu^{-1})\{u^2(u+v)^{-1}\}u^{-1}.$$

Hence, applying Lemma 2.3 twice, we see that $fgh^{-1} \in \mathcal{R}(W)$. Since $x \in U$ is arbitrary, it follows that $fgh^{-1} \in \mathcal{R}(U)$.

Corollary 2.7. If $1 \in \mathcal{R}(U)$, then $\mathcal{R}(U)$ is an algebra; and $f \in \mathcal{R}(U)$, $f > 0$ on U imply $f^{-1} \in \mathcal{R}(U)$.

Proposition 2.19. Suppose $1 \in \mathcal{R}(U)$. Let K be a compact set in U and let $\{U_j\}_{j=1}^n$ be a finite open covering of K. Then there exist $h_1, \ldots, h_n \in \mathcal{R}(U)$ such that $h_j \geq 0$ on U and Supp h_j is compact and contained in $U_j \cap U$ for each j and such that $\Sigma_{j=1}^n h_j = 1$ on K.

Proof. We can find a finite number of relatively compact open sets V_j, $j = 1, \ldots, n$, such that $\overline{V}_j \subset U_j \cap U$ for each j and $\bigcup_{j=1}^n V_j \supset K$. By Proposition 2.17 , there are $f_j \in \mathcal{R}(U)$, $j = 1, \ldots, n$, such that $f_j \geq 0$ on U, $f_j > 0$ on $\overline{V}_j$ and Supp f_j is compact and contained in $U_j \cap U$ for each j. Also, there is $g \in \mathcal{R}(U)$ such that $g \geq 0$ on U, $g = 1$ on K and Supp g is compact and contained in $\bigcup_{j=1}^n V_j$. Put

$$h_j = \begin{cases} g f_j \left(\sum_{j=1}^{n} f_j \right)^{-1} & \text{on } \bigcup_{j=1}^{n} V_j \\ 0 & \text{on } U \setminus \bigcup_{j=1}^{n} V_j . \end{cases}$$

Then, by Proposition 2.18 , we see that $h_j \in \mathcal{R}(U)$ for each j. It is easy to see that $h_1, \ldots, h_n$ satisfy the rest of required properties.

§3. Gradient measures

Throughout this section, let $(X, \mathcal{U})$ be a harmonic space and let $\mathcal{H} = \mathcal{H}_{\mathcal{U}}$.

3-1. Measure representations (cf. [26])

By a sheaf homomorphism σ of the sheaf $\mathcal{R}$ into the sheaf $\mathcal{M}$ of measures on X, we mean a set of mappings $\{\sigma_U\}_{U \in \mathcal{O}_X}$ such that

(i) for each $U \in \mathcal{O}_X$, σ_U is a linear mapping of $\mathcal{R}(U)$ into $\mathcal{M}(U)$,

(ii) if $U, U' \in \mathcal{O}_X$, $U' \subset U$ and $f \in \mathcal{R}(U)$, then

$$\sigma_U(f)|U' = \sigma_{U'}(f|U').$$

By virtue of property (ii), there will be no ambiguity in writing $\sigma(f)$ instead of $\sigma_U(f)$.

A sheaf homomorphism $\sigma: \mathcal{R} \to \mathcal{M}$ will be called a <u>measure representation</u> of $\mathcal{R}$ if it satisfies the following condition:

(iii) For each $U \in \mathcal{O}_X$ and $f \in \mathcal{R}(U)$, $\sigma(f) \geq 0$ on U if and only if f is superharmonic on U.

If σ is a measure representation of $\mathcal{R}$ and if $g \in \mathbf{C}(X)$ and $g > 0$ on X, then σ': $\sigma'(f) = g\sigma(f)$ for $f \in \mathcal{R}(U)$ defines another measure representation of $\mathcal{R}$.

Let W be an open set in X and $h \in \mathbf{C}(W)$, $h > 0$ on W. The sheaf considered with respect to the harmonic space $(W, \mathcal{U}_{W,h})$ will be denoted by $\mathcal{R}_W^{(h)}$, or simply by $\mathcal{R}^{(h)}$. Obviously, it is given by

$$\mathcal{R}^{(h)}(U) = \{f/h \mid f \in \mathcal{R}(U)\} \qquad \text{for } U \subset W.$$

Given a measure representation σ of $\mathcal{R}$,

$$\sigma^{(h)}(f) = \sigma(fh) \qquad \text{for } f \in \mathcal{R}^{(h)}(U),\ U \subset W$$

defines a measure representation of $\mathcal{R}_W^{(h)}$.

Example 3.1. Let $(X, \mathcal{H})$ be the Brelot's harmonic space given in Example 1.1 with $L=\Delta$, i.e., $X \subset \mathbb{R}^n$ and $\mathcal{H}(U) = \{u \in C^2(U) \mid \Delta u = 0\}$ for $U \in \mathcal{O}_X$. Then, for $f \in \mathcal{R}(U)$, Δf in the distribution sense is a measure on U. Letting $\sigma(f) = -\Delta f$, we obtain a measure representation (cf., e.g., [4], [15]).

In the case where L is a general elliptic linear differential operator as given in Example 1.1 , the existence of a measure representation σ of $\mathcal{R}$ is assured by a general theory given below (3-2.). Later (cf. Theorem 3.4 and Remark 3.5) we shall see that $C^2(U) \subset \mathcal{R}(U)$ and given a measure representation σ of $\mathcal{R}$, there corresponds a non-negative measure ν on X such that $\sigma(f) = (-Lf)\nu$ for $f \in C^2(U)$.

Example 3.2. Let $(X, \mathcal{H})$ be the Bauer's harmonic space given in Example 1.2 , i.e., $X \subset \mathbb{R}^{n+1}$ and $\mathcal{H}(U) = \{u \in C^2(U) \mid \frac{\partial u}{\partial x_{n+1}} = \Delta_n u\}$. In this case, for $f \in \mathcal{R}(U)$, $\sigma(f) = \frac{\partial f}{\partial x_{n+1}} - \Delta_n f$ exists in the distribution sense and σ defines a measure representation of $\mathcal{R}$ (cf. [33]).

Example 3.3. Let $(X, \mathcal{H})$ be the Bauer's harmonic space given in Example 1.3. Then, for $f \in \mathcal{R}(U)$, f'' in the distribution sense is a measure on $U \setminus \{0\}$ and the right-hand side derivative $f'_+(0)$ exists at 0 when $0 \in U$. Let

$$\varphi(t) = \begin{cases} -1 & \text{if } t > 0 \\ t & \text{if } t \leq 0. \end{cases}$$

Then we see that $\varphi f''$ is a measure on U and

$$\sigma(f) = \varphi f'' - f'_+(0)\varepsilon_0$$

defines a measure representation of $\mathcal{R}$.

Example 3.4. Let $(X, \mathcal{U})$ be the harmonic space given in Example 1.4. Then, for any $f \in \mathcal{R}(U)$, f' in the distribution sense is a measure on U and $\sigma(f) = -f'$ gives a measure representation σ of $\mathcal{R}$.

3-2. Existence of measure representations on Brelot's harmonic spaces (cf. [31])

In this subsection let $(X,\mathcal{H})$ be a Brelot's harmonic space and suppose it satisfies the following proportionality condition:

(Pr) For each P-domain U (P-domain = P-set which is a domain) and $y\in U$, if $p_1,p_2\in \mathcal{P}(U)$, p_1,p_2 are harmonic on $U\setminus\{y\}$ and $p_1 \not\equiv 0$, then there is a constant $\lambda \geq 0$ such that $p_2 = \lambda p_1$ on U.

The following theorem is shown in R.-M. Hervé [16] (also cf. [11; Theorem 11.5.2]):

Theorem of Hervé. Let $(X,\mathcal{H})$ be a Brelot's harmonic space satisfying (Pr) and let U be a P-domain in X. Then there exists a function (called a Green function on U) $G_U : U \times U \to]0,+\infty]$ satisfying the following two conditions:

(a) G_U is lower semicontinuous on $U \times U$ and is continuous on $U \times U \setminus \{(x,x)\,|\,x\in U\}$;

(b) for each $y\in U$, $G_U(\cdot,y)\in \mathcal{P}(U)$ and is harmonic on $U\setminus\{y\}$.

Furthermore, for any $p \in \mathcal{P}(U)$, there exists a unique non-negative measure μ on U such that

$$p(x) = \int_U G_U(x,y)\, d\mu(y) \quad , \quad x\in U.$$

A system $\{G_U\}_{U:\text{P-domain}}$ will be called a consistent-system of Green functions on X, if each G_U is a Green function on U and if, for any P-domains U, U' with $U' \subset U$ and for any $y\in U'$, there is $u_y\in \mathcal{H}(U')$ such that

$$G_U(x,y) = G_{U'}(x,y) + u_y(x) \quad , \quad x\in U'.$$

Lemma 3.1. Let $(X,\mathcal{H})$ be a Brelot's harmonic space satisfying (Pr) and let U_1,U_2 be two P-domains in X such that $U_1\cap U_2 \neq \emptyset$. If G_{U_1} and G_{U_2} are Green functions on U_1 and U_2, respectively, then there is $\lambda\in \mathcal{C}(U_1\cap U_2)$ such that $\lambda > 0$ on $U_1\cap U_2$ and

(3.1) $$G_{U_1}(\cdot,y) = \lambda(y)\, G_{U_2}(\cdot,y) + u_y$$

with $u_y\in \mathcal{H}(U_1\cap U_2)$ for each $y\in U_1\cap U_2$.

Proof. By Theorem 2.2 (Riesz decomposition theorem) and condition (Pr), we can write in the form (3.1) with $\lambda(y) > 0$ for each $y \in U_1 \cap U_2$. Thus, it is enough to prove that λ is continuous on $U_1 \cap U_2$. Let $y_o \in U_1 \cap U_2$ and choose a regular domain V such that $y_o \in V$ and $\overline{V} \subset U_1 \cap U_2$. Then, for each $y \in V$

$$G_{U_i}(\cdot,y) = p_{i,y} + H^V_{\varphi_{i,y}} \quad \text{with } p_{i,y} \in \mathcal{P}(V),\ i = 1,2,$$

where $\varphi_{i,y}(\xi) = G_{U_i}(\xi,y)$, $\xi \in \partial V$, $i = 1,2$. Then $p_{1,y} = \lambda(y)p_{2,y}$. Fix $x \in V\setminus\{y_o\}$. Then the mappings $y \mapsto G_{U_i}(x,y)$, $i = 1,2$ are continuous at y_o. Also, the mappings $y \mapsto G_{U_i}(\xi,y)$ are equi-continuous at y_o with respect to $\xi \in \partial V$, so that the mappings $y \mapsto H^V_{\varphi_{i,y}}(x)$ are also continuous at y_o ($i = 1,2$). Therefore, $y \mapsto p_{i,y}(x)$ are continuous at y_o, which implies that λ is continuous at y_o.

Proposition 3.1. If $(X,\mathcal{H})$ is a Brelot's harmonic space satisfying (Pr), then a consistent system of Green functions on X always exists.

Proof. Let $\{U_n\}$ be a locally finite covering of X by P-domains. Existence of such a covering is assured by Corollary 2.3 , Proposition 2.12 and the countability of X. For each n, let $\tilde{G}_{U_n}$ be a Green function on U_n. By the previous lemma, for any pair (i,j) such that $U_i \cap U_j \neq \emptyset$, there is $\lambda_{ij} \in \mathcal{C}(U_i \cap U_j)$ such that $\lambda_{ij} > 0$ on $U_i \cap_j$

and

$$(3.2) \qquad \tilde{G}_{U_i}(x,y) = \lambda_{ij}(y)\tilde{G}_{U_j}(x,y) + u_{ij,y}(x) \ , \quad x \in U_i \cap U_j$$

with $u_{ij,y} \in \mathcal{H}(U_i \cap U_j)$ for each $y \in U_i \cap U_j$. Then we can easily show that

$$(3.3) \qquad \lambda_{ij}\lambda_{ji} = 1 \quad \text{on } U_i \cap U_j \quad ; \qquad \lambda_{ij}\lambda_{jk}\lambda_{ki} = 1 \text{ on } U_i \cap U_j \cap U_k .$$

Let $\{\varphi_n\}$ be a partition of unity relative to $\{U_n\}$, i.e., $\varphi_n \in \mathcal{C}_o(X)$, $0 \leq \varphi_n \leq 1$, Supp $\varphi_n \subset U_n$ for each n and $\Sigma_n \varphi_n = 1$ on X. Let

$$\tilde{\lambda}_j^i = \begin{cases} \lambda_{ij}{}^{\varphi_i} & \text{on } U_i \cap U_j \\ 1 & \text{on } U_j \setminus (U_i \cap U_j) \end{cases}$$

for each i,j. Then $\tilde{\lambda}_j^i \in \mathcal{C}(U_j)$ and $\tilde{\lambda}_j^i > 0$ on U_j. Put

$$\lambda_j = \prod_i \tilde{\lambda}_j^i .$$

Since $\{U_n\}$ is locally finite, for any compact set K in U_j, $\tilde{\lambda}_j^i = 1$ on K except for a finite number of i, for a fixed j. Therefore, λ_j is well-defined and $\lambda_j \in \mathcal{C}(U_j)$, $\lambda_j > 0$ on U_j for each j. If $U_j \cap U_k \neq \emptyset$, then by (3.3) we have

$$\tilde{\lambda}_j^i \, \tilde{\lambda}_k^{i^{-1}} = \begin{cases} \lambda_{ij}{}^{\varphi_i} \lambda_{ik}{}^{-\varphi_i} = (\lambda_{ij}\lambda_{ik}{}^{-1})^{\varphi_i} = \lambda_{kj}{}^{\varphi_i} & \text{on } U_i \cap U_j \cap U_k \\ 1 & \text{on } U_j \cap U_k \setminus U_i , \end{cases}$$

so that $\lambda_j \lambda_k{}^{-1} = \lambda_{kj}$ on $U_j \cap U_k$. Now, let

$$G_{U_n}(x,y) = \lambda_n(y) \tilde{G}_{U_n}(x,y).$$

If $U_i \cap U_j \neq \emptyset$, then, by (3.2)

$$\begin{aligned} G_{U_i}(x,y) &= \lambda_i(y) \tilde{G}_{U_i}(x,y) \\ &= \lambda_i(y) \{\lambda_{ij}(y) \tilde{G}_{U_j}(x,y) + u_{ij,y}(x)\} \\ &= \lambda_j(y) \tilde{G}_{U_j}(x,y) + \lambda_i(y) u_{ij,y}(x) \\ &= G_{U_j}(x,y) + \lambda_i(y) u_{ij,y}(x) \end{aligned}$$

for $x,y \in U_i \cap U_j$.

Next, let U be any P-domain. Using Lemma 3.1 and the above result, we can find a Green function $G_U(x,y)$ on U such that

$$G_U(x,y) = G_{U_i}(x,y) + u_{i,y}(x) \quad , \qquad x,y \in U \cap U_i$$

with $u_{i,y} \in \mathcal{H}(U \cap U_i)$, for all i such that $U_i \cap U \neq \emptyset$. Then it is easy to see that $\{G_U\}_{U:\text{P-domain}}$ is a consistent system of Green functions.

If $\{G_U\}_{U:\text{P-domain}}$ is a consistent system of Green functions on X, then Theorem of Hervé implies that to each $f \in \mathcal{R}(U)$ there corresponds a unique measure $\sigma(f) \in \mathfrak{M}(U)$ such that

$$f|V = \int_V G_V(\cdot,y)\, d\sigma(f) + u_f^V$$

with $u_f^V \in \mathcal{H}(V)$ for every P-domain $V \subset U$ such that $f|V \in \mathcal{S}_C(V) - \mathcal{S}_C(V)$. Then, this σ defines a measure representation of $\mathcal{R}$. Thus, in view of Proposition 3.1 , there always exists a measure representation on a Brelot's harmonic space satisfying (Pr).

<u>Remark 3.1.</u> It is shown by K. Janssen [17] that a Green function on a P-set exists and integral representation of potentials as in the Theorem of Hervé is valid on a more general harmonic space. We can similarly show the existence of a measure representation in this case (see [26; §6, Proposition 5]).

<u>Remark 3.2.</u> If $\{G_U\}$ and $\{\tilde{G}_U\}$ are two consistent systems of Green functions on X, then there exists $\lambda \in \mathcal{C}(X)$ such that $\lambda > 0$ on X and $G_U(x,y) = \lambda(y)\tilde{G}_U(x,y)$ for all $x,y \in U$ and for all P-domains U. Thus, if σ and $\tilde{\sigma}$ are the corresponding measure representations, then $\tilde{\sigma}(f) = \lambda\sigma(f)$ for $f \in \mathcal{R}(U)$.

<u>3-3. Properties of measure representations</u> (cf. [28])

<u>Lemma 3.2.</u> Let $U \in \mathcal{O}_X$ and suppose $1 \in \mathcal{H}(U)$. Let $s_1, s_2 \in \mathcal{S}_C(U)$ and suppose $f = s_1 - s_2$ is bounded on U. Put $\alpha = \sup_U f$ and $\beta = \inf_U f$.

Then, for any $p \geq 1$.

$$v_p = -(\alpha-f)^p + p(\alpha-\beta)^{p-1}s_2$$

is superharmonic on U, i.e., $v_p \in \mathcal{S}_C(U)$.

<u>Proof.</u> Let $V \in \mathcal{O}_{rc}(U)$. Put $w_i = s_i - \mu^V s_i$ on V $(i = 1,2)$. Then, $w_i \geq 0$ on V $(i = 1,2)$. Since $1 \in \mathcal{H}(U)$ and $\alpha - f \geq 0$ on U, by Hölder's inequality we have

$$[\mu^V(\alpha-f)]^p \leq \mu^V[(\alpha-f)^p](\mu^V 1)^{p-1} = \mu^V[(\alpha-f)^p].$$

Hence,

$$\begin{aligned} \mu^V v_p &= -\mu^V[(\alpha-f)^p] + p(\alpha-\beta)^{p-1}\mu^V s_2 \\ &\leq -[\mu^V(\alpha-f)]^p + p(\alpha-\beta)^{p-1}\mu^V s_2 \\ &= -(\alpha-f+w_1-w_2)^p + p(\alpha-\beta)^{p-1}(s_2-w_2) \end{aligned}$$

on V. Since $\alpha-f \geq 0$ and $\alpha-f+w_1-w_2 = \mu^V(\alpha-f) \geq 0$, we have

$$(\alpha-f+w_1-w_2)^p \geq (\alpha-f)^p + p(w_1-w_2)(\alpha-f)^{p-1}$$

on V. Hence,

$$\begin{aligned} \mu^V v_p &\leq -(\alpha-f)^p - p(w_1-w_2)(\alpha-f)^{p-1} + p(\alpha-\beta)^{p-1}s_2 - p(\alpha-\beta)^{p-1}w_2 \\ &= v_p - pw_1(\alpha-f)^{p-1} + pw_2[(\alpha-f)^{p-1} - (\alpha-\beta)^{p-1}] \\ &\leq v_p \end{aligned}$$

on V. Therefore, v_p is superharmonic on U.

<u>Proposition 3.2.</u> Let $U \in \mathcal{O}_X$ and let p be an integer ≥ 2. If $1 \in \mathcal{H}(U)$ and σ is a measure representation of $\mathcal{R}$, then for any $f \in \mathcal{R}(U)$,

$$\sum_{k=1}^{p} (-1)^{k+1}\binom{p}{k} f^{p-k}\sigma(f^k) \geq 0 \qquad \text{on } U. \tag{3.4}$$

Proof. Let $V \in \mathcal{O}_{rc}(U)$ be such that $f|V = s_1 - s_2$ with $s_i \in \mathcal{S}_C(V)$, $i = 1,2$. Let $\varepsilon > 0$ ($\varepsilon < 1$) be given. Since f is continuous, for each $x \in V$ there is an open neighborhood W of x such that $W \subset V$ and

$$\sup_W f - \inf_W f < \varepsilon.$$

Put $\alpha = \sup_W f$ and $\beta = \inf_W f$. By the previous lemma, we have

$$-\sigma[(\alpha-f)^p] + p(\alpha-\beta)^{p-1}\sigma(s_2) \geq 0 \qquad \text{on } W.$$

Hence,

$$\sum_{k=1}^{p} (-1)^{k+1}\binom{p}{k}\alpha^{p-k}\sigma(f^k) \geq -p\varepsilon^{p-1}\sigma(s_2) \geq -p\varepsilon\,\sigma(s_2)$$

on W. Put $M = \sup_V |f|$ ($< +\infty$). Since

$$|\alpha^{p-k} - f^{p-k}| < (p-k)M^{p-k-1}\varepsilon \qquad \text{on } W$$

for each $k = 1,\ldots,p$, we have

$$\sum_{k=1}^{p} (-1)^{k+1}\binom{p}{k}f^{p-k}\sigma(f^k) \geq -\varepsilon[p\sigma(s_2) + p\sum_{k=1}^{p}\binom{p}{k-1}M^{p-k-1}|\sigma(f^k)|]$$

on W. The non-negative measure

$$\mu = p\sigma(s_2) + p\sum_{k=1}^{p}\binom{p}{k-1}M^{p-k-1}|\sigma(f^k)|$$

is defined on V and independent of W. Hence

$$\sum_{k=1}^{p} (-1)^{k+1}\binom{p}{k}f^{p-k}\sigma(f^k) \geq -\varepsilon\mu$$

on V. Since $\varepsilon > 0$ is arbitrary, we obtain (3.4) on V; and since such V's cover U, (3.4) holds on U.

Corollary 3.1. Under the same assumptions as in Proposition 3.2,

(3.5) $\sigma(fgh) - f\sigma(gh) - g\sigma(fh) - h\sigma(fg) + fg\sigma(h) + fh\sigma(g) + gh\sigma(f) = 0$

on U for any $f,g,h \in \mathcal{R}(U)$.

Proof. In case $p = 3$, the inequality (3.4) becomes

$$\sigma(f^3) - 3f\sigma(f^2) + 3f^2\sigma(f) \geq 0$$

for any $f \in \mathcal{R}(U)$. Applying this to $-f$, we obtain the converse inequality.
Hence,

$$\sigma(f^3) - 3f\sigma(f^2) + 3f^2\sigma(f) = 0 \tag{3.6}$$

for any $f \in \mathcal{R}(U)$. Now, given $f,g,h \in \mathcal{R}(U)$, consider the function $f+tg+sh$ with $t,s \in \mathbb{R}$ and apply (3.6) to this function. Using the linearity of σ, we see that the coefficient of ts is equal to six times the left hand side of (3.5). Therefore, we have (3.5).

Remark 3.3. As in the above proof, if p is odd (≥ 3), then equality holds in (3.4). It follows, as a matter of fact, that the equality holds in (3.4) for all $p \geq 3$. Then, we can show that

$$\sum_{k=0}^{p-1} (-1)^k \binom{p-1}{k} f^{(p-1)-k} \sigma(f^k g) = g \sum_{k=0}^{p-1} (-1)^k \binom{p-1}{k} f^{(p-1)-k} \sigma(f^k)$$

for $p \geq 3$, provided that $1 \in \mathcal{H}(U)$. Using this equality and considering the measure representation $\sigma^{(u)}$ of $\mathcal{R}_W^{(u)}$ on $W \subset U$ on which there is $u \in \mathcal{H}(W)$ with $u > 0$, even in case $1 \notin \mathcal{H}(U)$, if $1 \in \mathcal{R}(U)$ then we have

$$\sigma(f^2) - 2f\sigma(f) + f^2\sigma(1) \leq 0$$

and

$$\sum_{k=0}^{p} (-1)^{k+1} \binom{p}{k} f^{p-k} \sigma(f^k) = 0 \qquad \text{for } p = 3,4,\ldots$$

(cf. the proofs of Lemma 3.3 and Theorem 3.1 below).

3-4. Definition of gradient measures (cf. [26])

From now on, we assume that there exists a measure representation σ of $\mathcal{R}$ and fix it. We shall define gradient measures not for functions in $\mathcal{R}(U)$ but those in the following class:

$$\tilde{\mathcal{R}}(U) = \{f/h \mid f,h\in \mathcal{R}(U),\ h > 0 \text{ on } U\}, \qquad U\in \mathcal{O}_X.$$

By Proposition 2.18, we see that $\tilde{\mathcal{R}}(U)$ is an algebra containing constant functions. If we fix $h_o\in \mathcal{R}(X)$ such that $h_o > 0$ on X (cf. Corollary 2.5), then we may write

$$\tilde{\mathcal{R}}(U) = \{f/h_o \mid f\in \mathcal{R}(U)\},$$

again by Proposition 2.18. It follows that $\tilde{\mathcal{R}} : U \to \tilde{\mathcal{R}}(U)$ is a sheaf of functions on X. Also, by the same proposition, we see that $f\in \tilde{\mathcal{R}}(U)$ and $g\in \mathcal{R}(U)$ imply $fg\in \mathcal{R}(U)$. Obviously, $\tilde{\mathcal{R}}(U) = \mathcal{R}(U)$ if and only if $1\in \mathcal{R}(U)$.

Given $h\in \mathcal{C}(W)$ with $h > 0$ on W ($W\in \mathcal{O}_X$), the space $\tilde{\mathcal{R}}(U)$ corresponding to the harmonic space $(W, \mathcal{U}_{W,h})$ is given by

$$\tilde{\mathcal{R}}^{(h)}(U) = \{f/h \mid f\in \tilde{\mathcal{R}}(U)\}.$$

If $h\in \mathcal{R}(W)$, then $\tilde{\mathcal{R}}^{(h)}(U) = \mathcal{R}^{(h)}(U) = \tilde{\mathcal{R}}(U)$ for any $U \subset W$.

<u>Lemma 3.3.</u> Let $f,g\in \tilde{\mathcal{R}}(U)$, $h\in \mathcal{R}(U)$ and $h > 0$ on U. Then the signed measure

$$\frac{1}{2h}\{f\sigma(gh) + g\sigma(fh) - \sigma(fgh) - fg\sigma(h)\} \tag{3.7}$$

does not depend on h.

<u>Proof.</u> Let V be any open subset of U such that there is $u\in \mathcal{H}(V)$ with $u > 0$ on V. Since $1\in \mathcal{H}^{(u)}(V)$, Corollary 3.1 implies

$$\sigma^{(u)}(fgh/u) - f\sigma^{(u)}(gh/u) - g\sigma^{(u)}(fh/u) + fg\sigma^{(u)}(h/u)$$
$$= \frac{h}{u}\{\sigma^{(u)}(fg) - f\sigma^{(u)}(g) - g\sigma^{(u)}(f)\}$$

on V. Since $\sigma^{(u)}(f) = \sigma(uf)$ for any $f\in \tilde{\mathcal{R}}(V) = \mathcal{R}^{(u)}(V)$, it follows that

$$\frac{1}{2h}\{f\sigma(gh) + g\sigma(fh) - \sigma(fgh) - fg\sigma(h)\}$$
$$= \frac{1}{2u}\{f\sigma(gu) - g\sigma(fu) - \sigma(fgu)\},$$

the right hand side of which is independent of h. Hence, the measure (3.7) is independent of h on V. Since U is covered by such V's, we obtain the lemma.

We now define the <u>mutual gradient measure</u> $\delta_{[f,g]}$ of $f,g \in \tilde{\mathcal{R}}(U)$ (relative to σ) by (3.7), i.e.

(3.8) $$\delta_{[f,g]} = \frac{1}{2h}\{f\sigma(gh) + g\sigma(fh) - \sigma(fgh) - fg\sigma(h)\}$$

with some $h \in \mathcal{R}(U)$ which is strictly positive on U. If $1 \in \mathcal{R}(U)$, then we may take h = 1, and (3.8) becomes

(3.9) $$\delta_{[f,g]} = \frac{1}{2}\{f\sigma(g) + g\sigma(f) - \sigma(fg) - fg\sigma(1)\}.$$

The mapping $(f,g) \mapsto \delta_{[f,g]}$ is symmetric and bilinear on $\tilde{\mathcal{R}}(U) \times \tilde{\mathcal{R}}(U)$. The measure $\delta_f = \delta_{[f,f]}$ will be called the <u>gradient measure of</u> f $\in \tilde{\mathcal{R}}(U)$.

<u>Example 3.1'.</u> Let $(X, \mathcal{H})$ be the classical harmonic space, i.e., $X \subset \mathbb{R}^n$ and $\mathcal{H}(U) = \{u \in C^2(U) \mid \Delta u = 0\}$, and let $\sigma(f) = -\Delta f$ (in the distribution sense). Then for any $f \in \tilde{\mathcal{R}}(U) = \mathcal{R}(U)$, $\frac{\partial f}{\partial x_j}$, $j = 1,\dots,n$, exist as L^2_{loc}-functions on U and

$$\delta_{[f,g]} = \left(\sum_{j=1}^{n} \frac{\partial f}{\partial x_j}\frac{\partial g}{\partial x_j}\right) dx \qquad \text{for } f,g \in \mathcal{R}(U).$$

If $(X, \mathcal{H})$ is the Brelot's harmonic space given in Example 1.1 by an elliptic linear differential operator L, and if σ is a measure representation of $\mathcal{R}$ and $\sigma(f) = (-Lf)\nu$ for $f \in C^2(U)$ with a non-negative measure ν on X, then

$$\delta_{[f,g]} = \left(\sum_{i,j} a_{ij}\frac{\partial f}{\partial x_i}\frac{\partial g}{\partial x_j}\right)\nu \qquad \text{for } f,g \in C^2(U).$$

Example 3.2'. Let $(X,\mathcal{H})$ be the Bauer's harmonic space given in Example 1.2 and let $\sigma(f) = \frac{\partial f}{\partial x_{n+1}} - \Delta_n f$ for $f\in \tilde{\mathcal{R}}(U) = \mathcal{R}(U)$. Then

$$\delta_{[f,g]} = \left(\sum_{j=1}^{n} \frac{\partial f}{\partial x_j} \frac{\partial g}{\partial x_j} \right) dx_1 \dots dx_n dx_{n+1}$$

for $f,g\in \mathcal{R}(U)$.

Example 3.3'. Let $(X,\mathcal{H})$ be the Bauer's harmonic space given in Example 1.3 and let σ be given as in Example 3.3. Then, for $f,g\in \tilde{\mathcal{R}}(U) = \mathcal{R}(U)$, f', g' exist almost everywhere as L^2_{loc}-functions and

$$\delta_{[f,g]} = |\varphi| f'g' \, dt.$$

Example 3.4'. Let $(X,\mathcal{U})$ be the harmonic space given in Example 1.4 and let $\sigma(f) = -f'$ (in the distribution sense). Then, for any $f,g\in \tilde{\mathcal{R}}(U) = \mathcal{R}(U)$, $\delta_{[f,g]} = 0$.

3-5. Basic properties of gradient measures (cf. [26])

Theorem 3.1. $\delta_f \geq 0$ on U for any $f\in \tilde{\mathcal{R}}(U)$.

Proof. Let V be any open subset of U such that there is $u\in \mathcal{H}(V)$ with $u > 0$ on V. Applying Proposition 3.2 with $p = 2$ to the harmonic space $(V, \mathcal{U}_{V,u})$ and the measure representation $\sigma^{(u)}$ of $\mathcal{R}^{(u)}_V$, we obtain

$$2f\sigma^{(u)}(f) - \sigma^{(u)}(f^2) \geq 0$$

on V for $f\in \tilde{\mathcal{R}}(V) = \mathcal{R}^{(u)}(V)$. Hence

$$\delta_f = \frac{1}{2u} \{2f\sigma(fu) - \sigma(f^2 u)\} = \frac{1}{2u} \{2f\sigma^{(u)}(f) - \sigma^{(u)}(f^2)\} \geq 0$$

on V. Since such V's cover U, we see that $\delta_f \geq 0$ on U.

Proposition 3.3. (a) If $f,g\in \tilde{\mathcal{R}}(U)$ and $\lambda > 0$, then

$$|\delta_{[f,g]}| \leq \frac{1}{2} (\lambda \delta_f + \lambda^{-1} \delta_g).$$

(b) If $f, g \in \tilde{\mathcal{R}}(U)$ and A is a Borel subset of U, then

$$\{|\delta_{[f,g]}|(A)\}^2 \leq \delta_f(A)\delta_g(A).$$

(c) If $f \in \tilde{\mathcal{R}}(U)$ and $\delta_f = 0$ on U, then $\delta_{[f,g]} = 0$ on U for all $g \in \tilde{\mathcal{R}}(U)$.

<u>Proof.</u> For any $t \in \mathbb{R}$, by the above theorem,

$$0 \leq \delta_{f-tg} = \delta_f - 2t\delta_{[f,g]} + t^2\delta_g \quad \text{on } U.$$

Hence,

$$2t\delta_{[f,g]} \leq \delta_f + t^2\delta_g \quad \text{on } U.$$

It follows that

$$|\delta_{[f,g]}| \leq \frac{\delta_f + t^2\delta_g}{2|t|} \text{ on } U \quad \text{if } t \neq 0.$$

Hence, we have (a). If A is a relatively compact Borel set such that $\bar{A} \subset U$, then $\delta_f(A) < +\infty$, $\delta_g(A) < +\infty$ and

$$|\delta_{[f,g]}|(A) \leq \frac{1}{2}\{\lambda\delta_f(A) + \lambda^{-1}\delta_g(A)\} \quad (\lambda > 0).$$

If $\delta_f(A) = 0$ or $\delta_g(A) = 0$, then letting $\lambda \to \infty$ or $\lambda \to 0$, we see that $|\delta_{[f,g]}|(A) = 0$. If $\delta_f(A) \neq 0$ and $\delta_g(A) \neq 0$, then let $\lambda = \delta_f(A)^{-1/2}\delta_g(A)^{1/2}$. Then we obtain the inequality in (b). The case where A is any Borel set of U now immediately follows. (c) is an immediate consequence of (b).

<u>Proposition 3.4.</u> Let $W \in \mathcal{O}_X$, $h \in \mathcal{R}(W)$ and $h > 0$ on W. Then the mutual gradient measure $\delta^{(h)}_{[f,g]}$ of $f, g \in \mathcal{R}^{(h)}(U) = \tilde{\mathcal{R}}(U)$ $(U \subset W)$ with respect to the harmonic space $(W, \mathcal{U}_{W,h})$ and relative to the measure representation $\sigma^{(h)}$ is given by

$$\delta^{(h)}_{[f,g]} = h\delta_{[f,g]}.$$

Proof. Since $1 \in \mathcal{R}^{(h)}(U)$, we have

$$\delta^{(h)}_{[f,g]} = \frac{1}{2}\{f\sigma^{(h)}(g) + g\sigma^{(h)}(f) - \sigma^{(h)}(fg) - fg\sigma^{(h)}(1)\}$$

$$= \frac{1}{2}\{f\sigma(gh) + g\sigma(fh) - \sigma(fgh) - fg\sigma(h)\} = h\delta_{[f,g]}$$

Theorem 3.2. For $f,g,h \in \tilde{\mathcal{R}}(U)$.

(3.10) $$\delta_{[fg,h]} = f\delta_{[g,h]} + g\delta_{[f,h]}.$$

Proof. Let V be any open subset of U for which there is $u \in \mathcal{H}(V)$ with $u > 0$ on V. By using Corollary 3.1, we have

$$2\{\delta^{(u)}_{[fg,h]} - f\delta^{(u)}_{[g,h]} - g\delta^{(u)}_{[f,h]}\}$$

$$= fg\delta^{(u)}(h) + h\sigma^{(u)}(fg) - \sigma^{(u)}(fgh)$$

$$- f\{g\sigma^{(u)}(h) + h\sigma^{(u)}(g) - \sigma^{(u)}(gh)\}$$

$$- g\{f\sigma^{(u)}(h) + h\sigma^{(u)}(f) - \sigma^{(u)}(fh)\}$$

$$= -\{\sigma^{(u)}(fgh) - f\sigma^{(u)}(gh) - g\sigma^{(u)}(fh) - h\sigma^{(u)}(fg)$$

$$+ fg\sigma^{(u)}(h) + fh\sigma^{(u)}(g) + gh\sigma^{(u)}(f)\}$$

$$= 0$$

on V. Thus, in view of Proposition 3.4, (3.10) holds on V. Since U is covered by such V's, (3.10) holds on U.

Proposition 3.5. For $f \in \tilde{\mathcal{R}}(U)$,

$$\delta_{[f^+,f^-]} = 0, \text{ so that } \delta_{|f|} = \delta_f = \delta_{f^+} + \delta_{f^-} \text{ on } U,$$

where $f^+ = \max(f,0)$ and $f^- = \max(-f,0)$.

Proof. First we remark that $\tilde{\mathcal{R}}(U)$ is closed under max. and min. operations by virtue of Proposition 2.16. Let $f \in \tilde{\mathcal{R}}(U)$ and let $h \in \mathcal{R}(U)$ be strictly positive on U. Since $f^+f^- = 0$,

$$\delta_{[f^+,f^-]} = \frac{1}{2h} \{f^+\sigma(f^-h) + f^-\sigma(f^+h)\}.$$

The sets $U^+ = \{x \in U \mid f(x) > 0\}$ and $U^- = \{x \in U \mid f(x) < 0\}$ are open and $f^- = 0$ on U^+, $f^+ = 0$ on U^-. Hence $\sigma(f^-h) \mid U^+ = 0$ and $\sigma(f^+h)|U^- = 0$. Hence $\delta_{[f^+,f^-]} = 0$. Now,

$$\delta_{|f|} = \delta_{f^++f^-} = \delta_{f^+} + 2\delta_{[f^+,f^-]} + \delta_{f^-}$$

$$= \delta_{f^+} + \delta_{f^-}$$

$$= \delta_{f^+} - 2\delta_{[f^+,f^-]} + \delta_{f^-} = \delta_{f^+-f^-} = \delta_f.$$

Corollary 3.2. For $f,g \in \tilde{\mathcal{R}}(U)$,

$$\delta_{\max(f,g)} + \delta_{\min(f,g)} = \delta_f + \delta_g.$$

Proof. By the above proposition, we have

$$\delta_f + \delta_g - 2\delta_{[f,g]} = \delta_{f-g}$$

$$= \delta_{\max(f-g,0)} + \delta_{\min(f-g,0)}$$

$$= \delta_{\max(f,g)-g} + \delta_{\min(f,g)-g}$$

$$= \delta_{\max(f,g)} + \delta_{\min(f,g)} - 2\delta_{[f+g,g]} + 2\delta_g$$

$$= \delta_{\max(f,g)} + \delta_{\min(f,g)} - 2\delta_{[f,g]}.$$

3-6. Composition of functions in $\mathcal{R}(U)$ with C^2-functions (cf. [26])

<u>Lemma 3.4.</u> Let g_j, $f_n \in \mathcal{R}(U)$, $j = 1,\dots,m$; $n = 1,2,\dots$ and $\varphi_{j,n} \in \mathcal{C}(U)$, $j = 1,\dots,m$; $n = 1,2,\dots$. Suppose f_n converges to f locally uniformly on U and $\varphi_{j,n}$ converges to φ_j (as $n \to \infty$) locally uniformly on U for each j. Suppose, furthermore,

$$\sigma(f_n) = \sum_{j=1}^{m} \varphi_{j,n}\,\sigma(g_j)\,, \qquad n = 1,2,\dots .$$

Then $f \in \mathcal{R}(U)$ and

$$\sigma(f) = \sum_{j=1}^{m} \varphi_j \sigma(g_j).$$

<u>Proof.</u> Let V be any relatively compact open set such that $\overline{V} \subset U$ and $g_j|V \in \mathcal{S}_C(V)$ for all $j = 1,\dots,m$. Let $g_j|V = s_j^{(1)} - s_j^{(2)}$ with $s_j^{(k)} \in \mathcal{S}_C(V)$, $k = 1,2$; $j = 1,\dots,m$. Put

$$s = \sum_{j=1}^{m} \sum_{k=1}^{2} s_j^{(k)}.$$

Then, $|\sigma(g_j)| \leq \sigma(s)$ for all $j = 1,\dots,m$. Put

$$\gamma_n = \sum_{j=1}^{m} \sup_V |\varphi_{j,n} - \varphi_j|\,, \qquad n = 1,2,\dots .$$

By our assumption, $\gamma_n \to 0$ $(n \to \infty)$. Furthermore,

$$(3.11) \qquad \left|\sigma(f_n) - \sum_{j=1}^{m} \varphi_j \sigma(g_j)\right| = \left|\sum_{j=1}^{m} (\varphi_{j,n} - \varphi_j)\sigma(g_j)\right| \leq \gamma_n \sigma(s)$$

on V for all n. Let $\varepsilon > 0$ be arbitrarily given. Since φ_j are continuous, for each $x \in V$ there is an open neighborhood W_x of x such that $\overline{W}_x \subset V$ and

$$\sup_{W_x} \varphi_j - \inf_{W_x} \varphi_j < \varepsilon, \qquad j = 1,\dots,m.$$

Put $\alpha_j = \sup_{W_x} \varphi_j$ and $\beta_j = \inf_{W_x} \varphi_j$, $j = 1,\dots,m$; and put

$$v_1 = \sum_{j=1}^{m} \{\beta_j s - \alpha_j(s-g_j)\}, \quad v_2 = \sum_{j=1}^{m} \{\alpha_j s - \beta_j(s-g_j)\}.$$

Since $\sigma(s) \geq 0$ and $\sigma(s-g_j) \geq 0$ on V, we have

$$\begin{aligned} \sigma(v_1) &= \sum_{j=1}^{m} \{\beta_j\sigma(s) - \alpha_j\sigma(s-g_j)\} \\ &\leq \sum_{j=1}^{m} \{\varphi_j\sigma(s) - \varphi_j\sigma(s-g_j)\} = \sum_{j=1}^{m} \varphi_j\sigma(g_j) \\ &\leq \sum_{j=1}^{m} \{\alpha_j\sigma(s) - \beta_j\sigma(s-g_j)\} = \sigma(v_2) \end{aligned} \tag{3.12}$$

on W_x. From (3.11) and (3.12) it follows that

$$f_n - v_1 + \gamma_n s \in \mathcal{S}_C(W_x) \text{ and } -f_n + v_2 + \gamma_n s \in \mathcal{S}_C(W_x)$$

for all n. Since these functions converge uniformly on W_x, we conclude that

$$f - v_1 \in \mathcal{S}_C(W_x) \quad \text{and} \quad -f + v_2 \in \mathcal{S}_C(W_x). \tag{3.13}$$

Since $v_k \in \mathcal{S}_C(W_x) - \mathcal{S}_C(W_x)$, $k = 1,2$, it follows that $f \in \mathcal{S}_C(W_x) - \mathcal{S}_C(W_x)$. Since $x \in V$ is arbitrary, $f \in \mathcal{R}(V)$; and since such V's cover U, $f \in \mathcal{R}(U)$.

Furthermore, (3.13) shows that

$$\sigma(v_1) \leq \sigma(f) \leq \sigma(v_2) \quad \text{on } W_x.$$

Hence, in view of (3.12)

$$\left|\sigma(f) - \sum_{j=1}^{m} \varphi_j\sigma(g_j)\right| \leq \sigma(v_2) - \sigma(v_1) \leq 3\varepsilon m\sigma(s)$$

on W_x. The first and the last terms are independent of W_x.

Hence

$$|\sigma(f) - \sum_{j=1}^{m} \varphi_j \sigma(g_j)| \leq 3\varepsilon m\sigma(s) \quad \text{on } V.$$

Now, letting $\varepsilon \to 0$, we see that $\sigma(f) = \sum_{j=1}^{m} \varphi_j \sigma(g_j)$ on V, and hence on U.

Theorem 3.3. Let $f_1,\ldots,f_k \in \tilde{\mathcal{R}}(U)$ and put $\vec{f} = (f_1,\ldots,f_k)$. Let Ω be an open set in $\mathbb{R}^k$ containing $\vec{f}(U)$. If $\varphi \in C^2(\Omega)$, then $\varphi \circ \vec{f} \in \tilde{\mathcal{R}}(U)$ and the following equations hold:

(3.14)
$$\sum_{i,j=1}^{k} \left(\frac{\partial^2 \varphi}{\partial x_i \partial x_j} \circ \vec{f}\right) \delta_{[f_i,f_j]}$$
$$= \frac{1}{h}\left\{\sum_{j=1}^{k} \left(\frac{\partial \varphi}{\partial x_j} \circ \vec{f}\right)\{\sigma(f_j h) - f_j\sigma(h)\} + (\varphi\circ\vec{f})\sigma(h) - \sigma[(\varphi\circ\vec{f})h]\right\}$$

for any $h \in \mathcal{R}(U)$ such that $h > 0$ on U;

(3.15)
$$\delta_{[\varphi\circ\vec{f},g]} = \sum_{j=1}^{k} \left(\frac{\partial \varphi}{\partial x_j} \circ \vec{f}\right) \delta_{[f_j;g]}$$

for any $g \in \tilde{\mathcal{R}}(U)$.

Thus, for $\varphi, \psi \in C^2(\Omega)$, we have

(3.16)
$$\delta_{[\varphi\circ\vec{f},\psi\circ\vec{f}]} = \sum_{i,j=1}^{k} \left(\frac{\partial \varphi}{\partial x_i} \circ \vec{f}\right)\left(\frac{\partial \psi}{\partial x_j} \circ \vec{f}\right) \delta_{[f_i,f_j]}.$$

Proof. Let $\mathcal{a}$ be the set of all $\varphi \in C^2(\Omega)$ for which $\varphi\circ\vec{f} \in \tilde{\mathcal{R}}(U)$ and (3.14) and (3.15) hold.

(I) $1 \in \mathcal{a}$ and $x_j \in \mathcal{a}$, $j = 1,\ldots,k$. For, if $\varphi(x) \equiv 1$, then both sides of (3.14) and (3.15) reduce to zero; if $\varphi(x) = x_j$ ($x = (x_1,\ldots,x_k)$), then both sides of (3.14) are zero and both sides of (3.15) are equal to $\delta_{[f_j,g]}$.

(II) If $\varphi_1, \varphi_2 \in \mathcal{A}$, then $\varphi_1\varphi_2 \in \mathcal{A}$. To show this, let $\varphi = \varphi_1\varphi_2$. Note that $\varphi o \vec{f} \in \tilde{\mathcal{R}}(U)$, since $\tilde{\mathcal{R}}(U)$ is an algebra. Using Theorem 3.2, we have

$$\delta[\varphi o\vec{f},g] = \delta[(\varphi_1 o\vec{f})(\varphi_2 o\vec{f}),g]$$

$$= (\varphi_1 o\vec{f})\delta[\varphi_2 o\vec{f},g] + (\varphi_2 o\vec{f})\delta[\varphi_1 o\vec{f},g]$$

$$= (\varphi_1 o\vec{f})\sum_{j=1}^{k}\left(\frac{\partial\varphi_2}{\partial x_j} o\vec{f}\right)\delta[f_j,g] + (\varphi_2 o\vec{f})\sum_{j=1}^{k}\left(\frac{\partial\varphi_1}{\partial x_j} o\vec{f}\right)\delta[f_j,g]$$

$$= \sum_{j=1}^{k}\left(\frac{\partial\varphi}{\partial x_j} o\vec{f}\right)\delta[f_j,g],$$

so that (3.15) holds for φ. Next, noting that (3.16) is valid with $\varphi=\varphi_1$ and $\psi=\varphi_2$, we have

$$\sum_{i,j=1}^{k}\left(\frac{\partial^2\varphi}{\partial x_i\partial x_j} o\vec{f}\right)\delta[f_i,f_j]$$

$$= \sum_{i,j=1}^{k}(\varphi_1 of)\left(\frac{\partial^2\varphi_2}{\partial x_i\partial x_j} o\vec{f}\right)\delta[f_i,f_j] + \sum_{i,j=1}^{k}(\varphi_2 o\vec{f})\left(\frac{\partial^2\varphi_1}{\partial x_i\partial x_j} o\vec{f}\right)\delta[f_i,f_j]$$

$$+ 2\sum_{i,j=1}^{k}\left(\frac{\partial\varphi_1}{\partial x_i} o\vec{f}\right)\left(\frac{\partial\varphi_2}{\partial x_j} o\vec{f}\right)\delta[f_i,f_j]$$

$$= \frac{\varphi_1 o\vec{f}}{h}\left\{\sum_{j=1}^{k}\left(\frac{\partial\varphi_2}{\partial x_j} o\vec{f}\right)\{\sigma(f_j h) - f_j\sigma(h)\} + (\varphi_2 o\vec{f})\sigma(h) - \sigma[(\varphi_2 o\vec{f})h]\right\}$$

$$+ \frac{\varphi_2 o\vec{f}}{h}\left\{\sum_{j=1}^{k}\left(\frac{\partial\varphi_1}{\partial x_j} o\vec{f}\right)\{\sigma(f_j h)-f_j\sigma(h)\} + (\varphi_1 o\vec{f})\sigma(h) - \sigma[(\varphi_1 o\vec{f})h]\right\}$$

$$+ 2\delta[\varphi_1 o\vec{f},\varphi_2 o\vec{f}]$$

$$= \frac{1}{h}\{\sum_{j=1}^{k}(\frac{\partial\varphi}{\partial x_j}\circ\vec{f})\{\sigma(f_j h) - f_j\sigma(h)\} + 2(\varphi\circ\vec{f})\sigma(h)$$

$$- (\varphi_1\circ\vec{f})\sigma[(\varphi_2\circ\vec{f})h] - (\varphi_2\circ\vec{f})\sigma[(\varphi_1\circ\vec{f})h] + 2h\delta_{[\varphi_1\circ\vec{f},\varphi_2\circ\vec{f}]}\}$$

$$= \frac{1}{h}\{\sum_{j=1}^{k}(\frac{\partial\varphi}{\partial x_j}\circ\vec{f})\{\sigma(f_j h) - f_j\sigma(h)\} + (\varphi\circ\vec{f})\sigma(h) -$$

$$- \sigma[(\varphi\circ\vec{f})h]\}.$$

Thus, (3.14) holds for φ.

(III) By (I) and (II), we see that all polynomials in $x_1,\dots,x_k$ belong to $\mathcal{A}$. Let $\varphi \in C^2(\Omega)$. Then we can find $\varphi_n \in \mathcal{A}$ $(n = 1,2,\dots)$ such that

$$\varphi_n \to \varphi,\ \frac{\partial\varphi_n}{\partial x_j} \to \frac{\partial\varphi}{\partial x_j},\ \frac{\partial^2\varphi_n}{\partial x_i\partial x_j} \to \frac{\partial^2\varphi}{\partial x_i\partial x_j}\ (i,j = 1,\dots,k)$$

all locally uniformly on Ω.

By (3.14) for φ_n, we have

$$\sigma[(\varphi_n\circ f)h] = -h\sum_{i,j=1}^{k}(\frac{\partial^2\varphi_n}{\partial x_i\partial x_j}\circ\vec{f})\delta_{[f_i,f_j]}$$

$$+ \sum_{j=1}^{k}(\frac{\partial\varphi_n}{\partial x_j}\circ\vec{f})\{\sigma(f_j h) - f_j\sigma(h)\} +$$

$$+ (\varphi_n\circ\vec{f})\sigma(h),$$

$n = 1,2,\dots$. Since $\varphi_n\circ\vec{f} \to \varphi\circ\vec{f}$, $\frac{\partial\varphi_n}{\partial x_j}\circ\vec{f} \to \frac{\partial\varphi}{\partial x_j}\circ\vec{f}$,

$\frac{\partial^2\varphi_n}{\partial x_i\partial x_j}\circ\vec{f} \to \frac{\partial^2\varphi}{\partial x_i\partial x_j}\circ\vec{f}$ all locally uniformly on U as

$n \to \infty$, Lemma 3.4 implies that $(\varphi\circ\vec{f})h \in \mathcal{R}(U)$, i.e., $\varphi\circ\vec{f} \in \tilde{\mathcal{R}}(U)$, and

$$\sigma[(\varphi\circ\vec{f})h] = -h\sum_{i,j=1}^{k}(\frac{\partial^2\varphi}{\partial x_i\partial x_j}\circ\vec{f})\,\delta_{[f_j,f_j]}$$

$$+ \sum_{j=1}^{k}(\frac{\partial\varphi}{\partial x_j}\circ\vec{f})\{\sigma(f_j h) - f_j\sigma(h)\} + (\varphi\circ\vec{f})\sigma(h),$$

that is, (3.14) holds for φ.

(IV) Next, given $\varphi \in C^2(\Omega)$, consider the function

$$\Phi(x_1,\dots,x_k,x_{k+1}) = x_{k+1}\,\varphi(x_1,\dots,x_k).$$

Then $\Phi \in C^2(\Omega\times \mathbb{R})$. Given $g \in \tilde{\mathcal{R}}(U)$, applying (3.14) to $\vec{f}^* = (f_1,\dots,f_k,g)$ and Φ, we obtain

$$g\sum_{i,j=1}^{k}\left(\frac{\partial^2\varphi}{\partial x_i\partial x_j}\circ\vec{f}\right)\delta_{[f_i,f_j]} + 2\sum_{j=1}^{k}\left(\frac{\partial\varphi}{\partial x_j}\circ\vec{f}\right)\delta_{[f_j,g]}$$

$$= \frac{1}{h}\Big\{g\sum_{j=1}^{k}\left(\frac{\partial\varphi}{\partial x_j}\circ\vec{f}\right)\{\sigma(f_jh) - f_j\sigma(h)\} + (\varphi\circ f)\{\sigma(gh) - g\sigma(h)\}$$

$$+ (\varphi\circ\vec{f})g\sigma(h) - \sigma[(\varphi\circ\vec{f})gh]\Big\}.$$

Hence, again using (3.14) for $\vec{f}$ and φ, we have

$$2\sum_{j=1}^{k}\left(\frac{\partial\varphi}{\partial x_j}\circ\vec{f}\right)\delta_{[f_j,g]} = \frac{1}{h}\{(\varphi\circ\vec{f})\sigma(gh) + g\sigma[(\varphi\circ\vec{f})h] - \sigma[(\varphi\circ\vec{f})gh] - (\varphi\circ\vec{f})g\sigma(h)\}$$

$$= 2\delta_{[\varphi\circ\vec{f},g]},$$

which is the required equality (3.15).

<u>Remark 3.4.</u> Given $f \in \tilde{\mathcal{R}}(U)$, if $\varphi \in C^2(\Omega)$ with $f(U) \subset \Omega \subset \mathbb{R}^1$ and if $\varphi'' \neq 0$ on $f(U)$, then from (3.14) we deduce

(3.17) $$\delta_f = \frac{1}{(\varphi''\circ f)h}\,[(\varphi'\circ f)\{\sigma(fh) - f\sigma(h)\} - \{\sigma[(\varphi\circ f)h] - (\varphi\circ f)\sigma(h)\}].$$

In case $\varphi(t)=t^2$, this is nothing but the definition of δ_f (cf. (3.8)). Thus, we may define δ_f by (3.17) using any φ as above.

3-7. The case where X is a Euclidean domain (cf. [26])

As an application of Theorem 3.3 , we obtain the following result in the special case where X is an open set in $\mathbb{R}^k$:

Theorem 3.4. Suppose that the base space X of the given harmonic space $(X, \mathcal{U})$ is an open set in $\mathbb{R}^k$ $(k \geq 1)$ and suppose a measure representation σ of $\mathcal{R}$ is given. Assume that for each point $a \in X$ there is an open neighborhood V of a and a C^2-diffeomorphism $\Psi = (\psi_1, \ldots, \psi_k)$ of V onto an open set $V' \subset \mathbb{R}^k$ such that $\psi_j \in \tilde{\mathcal{R}}(V)$ for all $j = 1, \ldots, k$. (In particular, if all the coordinate functions x_j belong to $\mathcal{R}(X)$, then this assumption is fulfilled with Ψ = the identity mapping and $V = V' = X$). Then $C^2(U) \subset \tilde{\mathcal{R}}(U)$ for any $U \in \mathcal{O}_X$, and writing

$$\alpha_{ij} = \delta_{[x_i, x_j]}, \quad i,j = 1, \ldots, k,$$

we have

$$\delta_{[f,g]} = \sum_{i,j=1}^{k} \frac{\partial f}{\partial x_i} \frac{\partial g}{\partial x_j} \alpha_{ij} \tag{3.18}$$

for $f, g \in C^2(U)$. The matrix (α_{ij}) is symmetric, and

positive semidefinite in the sense that $\mu_\xi = \Sigma_{i,j}\, \xi_i \xi_j \alpha_{ij}$ is a non-negative measure on X for each $\xi = (\xi_1, \ldots, \xi_k) \in \mathbb{R}^k$.

If, in addition, $1 \in \mathcal{R}(X)$, then with $\beta_j = -\sigma(x_j) + x_j\sigma(1)$, $j = 1, \ldots, k$ and $\gamma = -\sigma(1)$, every $f \in C^2(U)$ satisfies the equation

$$\sum_{i,j=1}^{k} \frac{\partial^2 f}{\partial x_i \partial x_j} \alpha_{ij} + \sum_{j=1}^{k} \frac{\partial f}{\partial x_j} \beta_j + f\gamma = -\sigma(f) \tag{3.19}$$

on U.

Proof. Let $f \in C^2(U)$, $U \in \mathcal{O}_X$. For each $a \in U$, there is an open neighborhood V of a contained in U and a C^2-diffeomorphism $\Psi = (\psi_1, \ldots, \psi_k)$ of V onto $V' \subset \mathbb{R}^k$ such that $\psi_j \in \tilde{\mathcal{R}}(V)$ for all $j = 1, \ldots, k$.

Since $f\circ\Psi^{-1}\in C^2(V')$, Theorem 3.3 implies that $f|V = (f\circ\Psi^{-1})\circ\Psi\in\tilde{\mathcal{R}}(V)$. Hence $f\in\tilde{\mathcal{R}}(U)$, and so $C^2(U)\subset\tilde{\mathcal{R}}(U)$. Now, we obtain (3.18) and (3.19) from (3.15) and (3.14) in Theorem 3.3 , by considering $f_j(x) = x_j$, $j = 1,\ldots,k$. Obviously, (α_{ij}) is symmetric and if $\xi = (\xi_1,\ldots,\xi_k)\in\mathbb{R}^k$, then

$$\mu_\xi = \delta_{\Sigma_j\xi_j x_j} \geq 0$$

by Theorem 3.1.

Corollary 3.3. Under the same assumptions as in the above theorem (including the assumption $1\in\mathcal{R}(X)$), $u\in C^2(U)$ is harmonic on U if and only if

$$\sum_{i,j=1}^{k} \frac{\partial^2 u}{\partial x_i \partial x_j}\alpha_{ij} + \sum_{j=1}^{k}\frac{\partial u}{\partial x_j}\beta_j + u\gamma = 0$$

on U.

Remark 3.5. In Theorem 3.4 , if $(X, \mathcal{H}_u)$ is the harmonic space given in Example 1.1 , then we can show that there is a non-negative measure ν on X such that $\alpha_{ij} = a_{ij}\nu$, $\beta_j = b_j\nu$ and $\gamma = c\nu$, so that $q(f) = (-Lf)\nu$ for $f\in C^2(U)$ (cf. [27]).

Remark 3.6. Problems in the same direction as Theorem 3.4 are discussed in [3] and [30]; cf. [27].

PART II GREEN'S FORMULAE ON SELF-ADJOINT HARMONIC SPACES

In what follows, we shall restrict ourselves to Brelot's harmonic space satisfying propotionality condition (Pr) (see 3-2).

§4. Self-adjoint harmonic spaces and Green potentials

4-1. Dirichlet problems (cf. [2], [11])

Let $(X, \mathcal{H})$ be a Brelot's harmonic space and $\mathcal{U} = \mathcal{U}_{\mathcal{H}}$. Given $U \in \mathcal{O}_X$, let U^a denote the closure of U in the one point compactification X^a of X. (If X is compact, then $X^a = X$.) Thus, if U is relatively compact, then $U^a = \overline{U}$. We denote: $\partial^a U = U^a \setminus U$. We shall say that $U \in \mathcal{O}_X$ is an <u>MP^a-set</u> if $u \in \mathcal{U}(U)$ and

$$\lim_{x \to \xi, x \in U} u(x) \geq 0 \qquad \text{for all } \xi \in \partial^a U$$

imply $u \geq 0$ on U. An MP^a-set is an MP-set, but not vice versa. A relatively compact MP-set is an MP^a-set. Thus, a relatively compact open set whose closure is contained in a P-set is an MP^a-set by virtue of Proposition 2.15.

Given an MP^a-set U and a $[-\infty, +\infty]$-valued function φ on $\partial^a U$, we consider

$$\overline{\mathcal{U}}_\varphi^{U^a} = \left\{ u \in \mathcal{U}(U) \;\middle|\; \begin{array}{l} u \text{ is bounded below on } U, \\ \liminf_{x \to \xi, x \in U} u(x) \geq \varphi(\xi) \text{ for all } \xi \in \partial^a U \end{array} \right\}$$

and $\underline{\mathcal{U}}_\varphi^{U^a} = -\overline{\mathcal{U}}_{-\varphi}^{U^a}$. Put $\overline{H}_\varphi^{U^a} = \inf \overline{\mathcal{U}}_\varphi^{U^a}$ and $\underline{H}_\varphi^{U^a} = \sup \underline{\mathcal{U}}_\varphi^{U^a}$.

Then, $\underline{H}_\varphi^{U^a} \leq \overline{H}_\varphi^{U^a}$ and $\overline{H}_\varphi^{U^a}$ (or $\underline{H}_\varphi^{U^a}$) is harmonic on U^a if it assumes finite value at one point of U. φ is said to be <u>resolutive with respect to U^a</u> if $\underline{H}_\varphi^{U^a} = \overline{H}_\varphi^{U^a} \in \mathcal{X}(U)$. In this case we write $H_\varphi^{U^a}$ for $\underline{H}_\varphi^{U^a} = \overline{H}_\varphi^{U^a}$. If every $\varphi \in \mathbf{C}(\partial^a U)$ is resolutive with respect to U^a, then we say that U^a is resolutive. If U^a is resolutive, then for each $x \in U$, there is a non-negative measure ρ_x^U on $\partial^a U$ such that

$$\int \varphi \, d\rho_x^U = H_\varphi^{U^a}(x) \qquad \text{for all } \varphi \in C(\partial^a U).$$

Lemma 4.1. Let U be an MP^a-set.

(a) If φ,ψ are resolutive with respect to $\partial^a U$, then so are $\varphi+\psi$ and $c\varphi$ (c: const.) and

$$H^{U^a}_{\varphi+\psi} = H^{U^a}_{\varphi} + H^{U^a}_{\psi}, \qquad H^{U^a}_{c\varphi} = cH^{U^a}_{\varphi}.$$

(b) If $\{\varphi_n\}$ is a sequence of functions which are resolutive with respect to U^a, $\varphi_n \uparrow \varphi$ on $\partial^a U$ and $\underline{H}^{U^a}_{\varphi}$ is finite, then φ is resolutive with respect to U^a and $H^{U^a}_{\varphi_n} \uparrow H^{U^a}_{\varphi}$.

Proof. (a) It is easy to see that $\overline{H}^{U^a}_{\varphi+\psi} \leq \overline{H}^{U^a}_{\varphi} + \overline{H}^{U^a}_{\psi}$, which implies the resolutivity of $\varphi+\psi$ and the equality $H^{U^a}_{\varphi+\psi} = H^{U^a}_{\varphi} + H^{U^a}_{\psi}$. If $c \geq 0$, then $\overline{H}^{U^a}_{c\varphi} = c\overline{H}^{U^a}_{\varphi}$; if $c \leq 0$, then $\overline{H}^{U^a}_{c\varphi} = c\underline{H}^{U^a}_{\varphi}$. Hence $c\varphi$ is resolutive and $H^{U^a}_{c\varphi} = cH^{U^a}_{\varphi}$.

(b) Obviously, $H^{U^a}_{\varphi_n} \uparrow$ $(n\uparrow\infty)$. Let $u = \lim_{n\to\infty} H^{U^a}_{\varphi_n}$. Since $H^{U^a}_{\varphi_n} \leq \underline{H}^{U^a}_{\varphi}$ and $\underline{H}^{U^a}_{\varphi}$ is finite, $u \in \mathcal{H}(U)$ and $u \leq \underline{H}^{U^a}_{\varphi}$.

Fix $x \in U$. For each $\varepsilon > 0$ and for each n, there is $s_n \in \overline{\mathcal{U}}^{U^a}_{\varphi_n}$ such that $s_n(x) \leq H^{U^a}_{\varphi_n}(x) + 2^{-n}\varepsilon$. Put $s = u + \Sigma^{\infty}_{n=1}(s_n - H^{U^a}_{\varphi_n})$.

Then $s \in \mathcal{U}(U)$ and $s(x) \leq u(x) + \varepsilon$. Furthermore, since $s \geq u + (s_n - H^{U^a}_{\varphi_n}) \geq s_n$ for all n, $\liminf_{x\to\xi} s(x) \geq \varphi_n(\xi)$ for all $\xi \in \partial^a U$ and for all n, so that $s \in \overline{\mathcal{U}}^{U^a}_{\varphi}$. Hence $s \geq \overline{H}^{U^a}_{\varphi}$, i.e., $u(x)+\varepsilon \geq \overline{H}^{U^a}_{\varphi}(x)$. Since $\varepsilon > o$ and $x \in U$ are arbitrary, we have $u \geq \overline{H}^{U^a}_{\varphi}$. Therefore, φ is resolutive with respect to U^a and $H^{U^a}_{\varphi_n} \uparrow H^{U^a}_{\varphi}$.

Corollary 4.1. If U^a is resolutive, then any bounded lower semicontinuous function φ on $\partial^a U$ is resolutive and

$$H^{U^a}_{\varphi}(x) = \int \varphi \, d\rho^U_x \qquad \text{for all } x \in U.$$

Proof. We can choose $\varphi_n \in \mathcal{C}(\partial^a U)$ such that $\varphi_n \uparrow \varphi$. Then, by (b) of the previous lemma, φ is resolutive with respect to U^a and

$$H_\varphi^{U^a}(x) = \lim_{n\to\infty} H_{\varphi_n}^{U^a}(x) = \lim_{n\to\infty} \int \varphi_n \, d\rho_x^U = \int \varphi \, d\rho_x^U.$$

Lemma 4.2. If there is $s \in \mathcal{S}_C(U)$ such that $\inf_U s > 0$, then U is an MP^a-set. In this case, if $\{\varphi_n\}$ is a sequence of functions which are resolutive with respect to U^a and $\varphi_n \to \varphi$ uniformly on $\partial^a U$, then φ is resolutive with respect to U^a and $H_{\varphi_n}^{U^a} \to H_\varphi^{U^a}$ on U.

Proof. By the same arguments as in the proof of Proposition 1.2, we see that U is an MP^a-set. Given $\{\varphi_n\}$ as in the lemma, let $M_n = \sup_{\partial^a U} |\varphi_n - \varphi|$. Let $\alpha = \inf_U s$. If $u \in \overline{\mathcal{U}}_\varphi^{U^a}$, then $u + M_n \alpha^{-1} s \in \overline{\mathcal{U}}_{\varphi_n}^{U^a}$, so that $\overline{H}_\varphi^{U^a} + M_n \alpha^{-1} s \geq H_{\varphi_n}^{U^a}$. Similarly, $\underline{H}_\varphi^{U^a} - M_n \alpha^{-1} s \leq H_{\varphi_n}^{U^a}$. Hence $0 \leq \overline{H}_\varphi^{U^a} - \underline{H}_\varphi^{U^a} \leq 2M_n \alpha^{-1} s \to 0$ $(n\to\infty)$. Hence φ is resolutive with respect to U^a. Also, $|H_\varphi^{U^a} - H_{\varphi_n}^{U^a}| \leq M_n \alpha^{-1} s \to 0$ $(n\to\infty)$.

Let U be a P-set and let $p \in \mathcal{P}(U)$. A potential e_p on U is called an Evans function for p if for any $\varepsilon > 0$ there is a compact set K_ε in U such that $p \leq \varepsilon e_p$ on $U \setminus K_\varepsilon$.

Lemma 4.3. Let U be a P-set and $p \in \mathcal{P}(U)$. Then an Evans function for p always exists.

Proof. Let $\{W_n\}$ be an exhaustion of U and put

$$p_n = R_U(\chi_{U\setminus \overline{W}_n} p), \qquad n = 1, 2, \ldots .$$

Then, $p_n \in \mathcal{P}(U)$, p_n is harmonic on W_n, $p_n \le p$ and $\{p_n\}$ is decreasing. Since $u = \lim_{n\to\infty} p_n$ is harmonic on U and $0 \le u \le p$, $u = 0$. It follows that $p_n \to 0$ locally uniformly on U. Hence, we can choose a subsequence $\{p_{n_k}\}$ such that

$$e_p = \sum_{k=1}^{\infty} p_{n_k}$$

converges locally uniformly on U. Then $e_p \in \mathcal{P}(U)$ by Proposition 2. Since $p_n = p$ on $U \setminus \overline{W}_n$, given $\varepsilon > 0$, if we choose an integer ℓ so that $\varepsilon\,\ell \ge 1$, then

$$\varepsilon e_p(x) \ge \varepsilon \sum_{k=1}^{\ell} p_{n_k}(x) = \varepsilon \ell p(x) \ge p(x)$$

for $x \in U \setminus \overline{W}_{n_\ell}$. Therefore e_p is an Evans function for p.

<u>Lemma 4.4.</u> Let U' be a P-set and $U \subset U'$ be an MP^a-set. Let $p \in \mathcal{P}(U')$ and put

$$\psi_p = \begin{cases} p & \text{on } \partial U \cap U' \\ 0 & \text{on } \partial^a U \setminus U'. \end{cases}$$

Then, ψ_p is resolutive with respect to U^a and $H^{U^a}_{\psi_p}$ is the greatest harmonic minorant of p on U.

<u>Proof.</u> Obviously, $p \in \overline{\mathcal{U}}^{U^a}_{\psi_p}$. Hence $0 \le \overline{H}^{U^a}_{\psi_p} \le p$, which also implies that $\overline{H}^{U^a}_{\psi_p} \in \mathcal{H}(U)$. Let e_p be an Evans function for p. If u is the greatest harmonic minorant of p on U, then $\overline{H}^{U^a}_{\psi_p} \le u \le p$ on U. Hence, for any $\varepsilon > 0$, $u - \varepsilon e_p \in \underline{\mathcal{U}}^{U^a}_{\psi_p}$, so that

$u - \varepsilon e_p \le \underline{H}^{U^a}_{\psi_p}$.

Therefore, $u \le \underline{H}^{U^a}_{\psi_p}$, which implies that ψ_p is resolutive with respect to U^a and $H^{U^a}_{\psi_p} = u$.

Given $U \in \mathcal{O}_X$ and $A \subset \partial^a U$, a non-negative superharmonic function s on U is called an Evans function on U for A if

$$\lim_{x\in U, x\to\xi} s(x) = +\infty$$

for all $\xi \in A$. If $A=\emptyset$, then any non-negative superharmonic function on U is an Evans function for A.

Proposition 4.1. Let U' be a P-set and U be an open subset of U' such that

(a) there is $s_o \in \mathcal{S}(U)$ such that $\inf_U s_o > o$, and

(b) there is an Evans function on U for the set $\partial^a U \setminus U'$.

Then U^a is resolutive and $\rho_x^U(\partial^a U \setminus U') = 0$ for all $x \in U$. In particular, if U is a relatively compact open set such that $\overline{U}$ is contained in a P-set, then $\overline{U} = U^a$ is resolutive (in this case $\rho_x^U = \mu_x^U$ in the notation in 1-1.).

Proof. Let $\tilde{s}$ be an Evans function on U for $\partial^a U \setminus U'$. Let $\varphi \in \mathcal{C}(\partial^a U)$ be given. For any $\varepsilon > 0$, there is a compact set K_ε in U' such that

$$\varepsilon \tilde{s} \geq \sup_{\partial^a U} |\varphi| \qquad \text{on } U \setminus K_\varepsilon .$$

Choose $f_\varepsilon \in \mathcal{C}_o(U')$ such that $0 \leq f_\varepsilon \leq 1$ on U' and $f_\varepsilon = 1$ on K_ε.

By Theorem 2.3 , there are $p_n, q_n \in \mathcal{P}(U') \cap \mathcal{C}(U')$ such that Supp $(p_n - q_n)$ is compact in U' and

$$|f_\varepsilon \varphi - (p_n - q_n)| < \frac{1}{n} \qquad \text{on } \partial U \cap U' .$$

By the previous lemma, ψ_{p_n}, ψ_{q_n} are resolutive with respect to U^a.

Hence, $\psi_{p_n} - \psi_{q_n}$ is resolutive with respect to U^a by Lemma 4.1.

Let

$$\varphi_\varepsilon = \begin{cases} f_\varepsilon \varphi & \text{on } \partial U \cap U' \\ 0 & \text{on } \partial^a U \setminus U'. \end{cases}$$

Then, $\psi_{p_n} - \psi_{q_n} \to \varphi_\varepsilon$ uniformly on $\partial^a U$ as $n \to \infty$, so that by Lemma 4.2., φ_ε is resolutive with respect to U^a.

For any $u \in \overline{\mathcal{U}}^{U^a}_{\varphi_\varepsilon}$, $u + \varepsilon\,\tilde{s} \in \overline{\mathcal{U}}^{U^a}_{\varphi}$. Hence $H^{U^a}_{\varphi_\varepsilon} + \varepsilon\,\tilde{s} \geq \overline{H}^{U^a}_{\varphi}$.

Similarly, we have $H^{U^a}_{\varphi_\varepsilon} - \varepsilon\,\tilde{s} \leq \underline{H}^{U^a}_{\varphi}$. Therefore

$$0 \leq \overline{H}^{U^a}_{\varphi} - \underline{H}^{U^a}_{\varphi} \leq 2\varepsilon\,\tilde{s} \qquad \text{on } U.$$

Since $\varepsilon > 0$ is arbitrary, we conclude that φ is resolutive with respect to U^a.

Next, given $\varepsilon > 0$, choose a compact set K'_ε in U' such that $s(x) \geq 1/\varepsilon$ for $x \in U \setminus K'_\varepsilon$. If $\varphi \in \mathcal{C}(\partial^a U)$ satisfies $\varphi = 1$ on $\partial^a U \setminus U'$, supp $\varphi \subset \partial^a U \setminus K'_\varepsilon$ and $0 \leq \varphi \leq 1$ on $\partial^a U$, then $\varepsilon\,\tilde{s} \in \overline{\mathcal{U}}^{U^a}_{\varphi}$, so that

$$\varepsilon\,\tilde{s}(x) \geq H^{U^a}_{\varphi}(x) \geq \rho^U_x\,(\partial^a U \setminus U') \qquad \text{for } x \in U.$$

Since $\varepsilon > 0$ is arbitrary, it follows that $\rho^U_x(\partial^a U \setminus U') = 0$ for all $x \in U$.

4-2. Symmetric Green functions (cf. [24],[22])

In this subsection, let $(X, \mathcal{H})$ be a Brelot's harmonic space and assume that it satisfies (Pr).

Given a P-domain U in X, a Green function G_U for U is called symmetric if $G_U(x,y) = G_U(y,x)$ for all $x,y \in U$.

<u>Proposition 4.2.</u> Let U' be a P-domain in X and suppose there exists a symmetric Green function $G_{U'}$ for U'. Let U be an open subset of U' such that U^a is resolutive. (In particular, this assumption is satisfied if U is relatively compact and $\overline{U} \subset U'$.)
For $x \in U$, put

$$\Psi_x(\xi) = \begin{cases} G_{U'}(x,\xi) & \text{if } \xi \in \partial U \cap U' \\ 0 & \text{if } \xi \in \partial^a U \setminus U'. \end{cases}$$

Then, Ψ_x is resolutive with respect to U^a for each $x \in U$ and

$$H^{U^a}_{\Psi_x}(y) = H^{U^a}_{\Psi_y}(x) \qquad \text{for all } x,y \in U.$$

Furthermore, $G_U(x,y) = G_{U'}(x,y) - H^{U^a}_{\Psi_x}(y) \quad (x,y \in U)$ gives a symmetric Green function for U, in case U is a domain.

<u>Proof.</u> By Lemma 4.4 , Ψ_x is resolutive with respect to U^a and $H^{U^a}_{\Psi_x} \leq G_{U'}(x,\cdot)$ for each $x \in U$. For $y \in U$, put

$$w_y(x) = H^{U^a}_{\Psi_x}(y) = \int_{\partial U \cap U'} G_{U'}(x,\xi)\, d\rho_y^{U}(\xi), \quad x \in U.$$

Since $x \to G_{U'}(x,\xi)$ is harmonic on U for each $\xi \in \partial U \cap U'$, we see that w_y is harmonic on U. Let e_y be an Evans function on U' for $G_{U'}(\cdot,y)$. Since $w_y \leq G_{U'}(\cdot,y)$, $w_y - \varepsilon e_y \in \underline{\mathcal{U}}^{U^a}_{\Psi_y}$ for any $\varepsilon > 0$. Hence $w_y \leq H^{U^a}_{\Psi_y}$, i.e.,

$$H^{U^a}_{\Psi_x}(y) \leq H^{U^a}_{\Psi_y}(x).$$

Since this is true for any $x,y \in U$, the equality holds. Then,

$$G_U(x,y) = G_{U'}(x,y) - H^{U^a}_{\Psi_y}(x) \qquad (x,y \in U)$$

is symmetric. Since $H_{\Psi_y}^{U^a}$ is the greatest harmonic minorant of $G_{U'}(\cdot,y)$ on U by Lemma 4.4 , $G_U(\cdot,y)\in \mathcal{P}(U)$ for each y. It then follows that $G_U(x,y)$ is a Green function for U.

Proposition 4.3. Let U' be a P-domain, U be a subdomain of U' and suppose both U' and U possess symmetric Green functions $G_{U'}$ and G_U, respectively. Then there exists a constant $\lambda > 0$ such that

$$G_{U'}(x,y) = \lambda\, G_U(x,y) + h_y(x), \quad x,y\in U$$

with $h_y \in \mathcal{H}(U)$ for each $y\in U$.

Proof. By (Pr), for each $y\in U$, there is $\lambda(y) > 0$ such that

(4.1) $$G_{U'}(x,y) = \lambda(y)G_U(x,y) + h_y(x) \quad , x\in U$$

with $h_y\in \mathcal{H}(U)$. Let V be any relatively compact domain such that $\overline{V} \subset U$. By the previous proposition,

$$G_{U'}(x,y) = G_V(x,y) + u_y(x) \quad , x\in V$$

$$G_U(x,y) = \tilde{G}_V(x,y) + \tilde{u}_y(x) \quad , x\in V$$

for each $y\in V$ with u_y, $\tilde{u}_y \in \mathcal{H}(V)$ and G_V, $\tilde{G}_V$ being symmetric Green functions on V. Then, in view of (4.1), we have

$$G_V(x,y) = \lambda(y)\, \tilde{G}_V(x,y)$$

for all $x,y\in V$. Since both G_V and $\tilde{G}_V$ are symmetric, $\lambda(x) = \lambda(y)$ for all $x,y\in V$, i.e., λ is constant on V. It follows that λ is constant on U and the proposition is proved.

4-3. Self-adjoint harmonic spaces (cf. [24])

A Brelot's harmonic space $(X,\mathcal{H})$ satisfying (Pr) is called a self-adjoint harmonic space if there exists a consistent system $\{G_U\}_{U:P\text{-domain}}$ of symmetric Green functions. Such a system, if exists, is uniquely determined up to a multiplicative constant independent of U on each connected component of X.

Example 4.1. Let X be an open set in $\mathbb{R}^n$ and let

$$Lf = \sum_{i,j=1}^{n} \frac{\partial}{\partial x_i} \left(a_{ij} \frac{\partial f}{\partial x_j}\right) + cf,$$

where a_{ij}, $i,j = 1,\ldots,n$ are C^1-functions on X whose first order partial derivatives are locally Hölder continuous on X and c is a locally Hölder continuous function on X. For $U \in \mathcal{O}_X$ set

$$\mathcal{H}(U) = \{u \in C^2(U) \mid Lu = 0\}.$$

Then, $(X, \mathcal{H})$ is a self-adjoint harmonic space (see [16]). In particular, the classical case, i.e., the case $L = \Delta$, is self-adjoint.

Let $(X, \mathcal{H})$ be a Brelot's harmonic space satisfying (Pr). If X itself is a P-domain and possesses a symmetric Green function, then we see by Proposition 4.2 that $(X, \mathcal{H})$ is self-adjoint. More generally, we have

Proposition 4.4. Let $(X, \mathcal{H})$ be a Brelot's harmonic space satisfying (Pr). Suppose there exists an increasing sequence $\{U_n\}$ of P-domains such that each U_n possesses a symmetric Green function and $\bigcup_n U_n = X$. Then $(X, \mathcal{H})$ is self-adjoint.

Proof. By using Proposition 4.3, we can find symmetric Green functions G_{U_n} for U_n, $n = 1,2,\ldots$, such that if $n<m$ then

$$G_{U_m}(x,y) = G_{U_n}(x,y) + h_{m,n}(x,y) \tag{4.2}$$

for $x,y \in U_n$ with $h_{m,n}(\cdot,y) \in \mathcal{H}(U_n)$. If U is a relatively compact domain, then choose n such that $\overline{U} \subset U_n$. By Proposition 4.2, we have a symmetric Green function G_U on U such that

$$G_{U_n}(x,y) = G_U(x,y) + h_n(x,y)$$

for $x,y \in U$ with $h_n(\cdot,y) \in \mathcal{H}(U)$. By virtue of (4.2), G_U does not depend on the choice of n. Now, let U be any P-domain. Let $\{W_n\}$ be an exhaustion of U. For each n, G_{W_n} is already defined. It is easy to see that $\{G_{W_n}\}$ is monotone increasing. Since U is a P-domain, we see that

$$G_U(x,y) = \lim_{n\to\infty} G_{W_n}(x,y)$$

is a symmetric Green function on U. In fact, if $\tilde{G}_U$ is some (not necessarily symmetric) Green function on U, then

$$\tilde{G}_U(x,y) = \lambda(y)G_{W_n}(x,y) + \tilde{h}_{n,y}(x), \quad x\in W_n$$

with $\tilde{h}_{n,y}\in \mathcal{H}(W_n)$ and $\lambda(y) > 0$ for each $y\in W_n$, which implies that $\lambda(y)G_U(x,y) \leq \tilde{G}_U(x,y)$ for all $x,y\in U$, and hence $G_U(\cdot,y)$ is a potential on U. From our construction of G_U, it is easy to see that $\{G_U\}_{U:\text{P-domain}}$ is a consistent system.

Example 4.2. Let X be an open interval in $\mathbb{R}$. Let ω_1,ω_2 be continuous functions on X such that $\omega_1^2+\omega_2^2 > 0$ everywhere on X, ω_2/ω_1 is strictly increasing on any interval $U \subset X$ on which $\omega_1 \neq 0$ and ω_1/ω_2 is strictly decreasing on any interval $U' \subset X$ on which $\omega_2 \neq 0$. (A typical example of such a pair (ω_1,ω_2) is $\omega_1 = 1$ and ω_2: strictly increasing; another example is $\omega_1(t) = \cos t$ and $\omega_2(t) = \sin t$). Put

$$\mathcal{H}(U) = \{\alpha\omega_1+\beta\omega_2 \mid \alpha,\beta\in \mathbb{R}\}$$

for an open interval $U \subset X$ and

$$\mathcal{H}(U) = \{u\in\mathcal{C}(U) \mid u|_{U'}\in \mathcal{H}(U') \text{ for any component } U' \text{ of } U\}$$

for $U\in \mathcal{O}_X$. Then $(X, \mathcal{H})$ is a Brelot's harmonic space satisfying (Pr). (In fact, we can show that any Brelot's harmonic structure on X is given in this way.)

If $U =]a,b[$ with $a,b \in X$, then

$$U\text{:P-domain} \iff \begin{vmatrix} \omega_1(x), \omega_2(x) \\ \omega_1(y), \omega_2(y) \end{vmatrix} > 0 \text{ whenever } a \leq x < y \leq b.$$

In this case, U has a symmetric Green function

$$G_U(x,y) = \frac{\{\omega_1(a)\omega_2(x)-\omega_2(a)\omega_1(x)\}\{\omega_1(y)\omega_2(b)-\omega_2(y)\omega_1(b)\}}{\omega_1(a)\omega_2(b)-\omega_2(a)\omega_1(b)}, \quad \text{if } x \leq y$$

$$G_U(y,x) = G_U(x,y), \qquad \text{if } x \geq y.$$

We see that $\{G_U\}$ is consistent and $(X, \mathcal{H})$ is self-adjoint (cf. the arguments at the end of the proof of Proposition 4.4).

Remark 4.1. Let $(X, \mathcal{H})$ be a Brelot's harmonic space satisfying (Pr). Even if each point $x \in X$ has a neighborhood which is a P-domain possessing a symmetric Green function, $(X, \mathcal{H})$ may not be self-adjoint. For example, let X be the one-dimensional torus: $X = \{e^{i\theta} \mid \theta \in \mathbb{R}\}$ and for $U = \{e^{i\theta} \mid \alpha < \theta < \beta\}$ with $-2\pi < \alpha < \beta < 2\pi$, $\beta-\alpha < 2\pi$, let

$$\mathcal{H}(U) = \begin{cases} \{u \in \mathcal{C}(U) \mid u \text{ is linear in } \theta\} & \text{if } 1 \notin U \\ \left\{u \in \mathcal{C}(U) \;\middle|\; \begin{array}{l} u \text{ is linear in } \theta \text{ for } \alpha<\theta<0 \text{ and } 0<\theta<\beta, \\ u(1) = \frac{1}{3}\{2u(e^{i\varepsilon}) + u(e^{-i\varepsilon})\} \text{ for small } \varepsilon>0 \end{array}\right\} & \text{if } 1 \in U. \end{cases}$$

From these $\mathcal{H}(U)$'s we can define a sheaf $\mathcal{H}$ on X. Then $(X, \mathcal{H})$ is a Brelot's harmonic space satisfying (Pr) and both $U_1 = \{e^{i\theta} \mid 0 < \theta < 2\pi\}$ and $U_2 = \{e^{i\theta} \mid -\pi < \theta < \pi\}$ have symmetric Green functions (cf. Example 4.2); but we cannot have a consistent system of symmetric Green functions.

Proposition 4.5. Let $(X, \mathcal{H})$ be a self-adjoint harmonic space, $W \in \mathcal{O}_X$ and $h \in \mathcal{C}(W)$ be strictly positive on W. Then $(W, \mathcal{H}_{W,h})$, where $\mathcal{H}_{W,h}(U) = \{u/h \mid u \in \mathcal{H}(U)\}$ for $U \in \mathcal{O}_W$, is again self-adjoint.

<u>Proof.</u> Let $\{G_U\}_{U:P\text{-domain}}$ be a consistent system of symmetric Green functions on X. For any P-domain $U \subset W$, let

$$(4.3) \qquad G_U^{(h)}(x,y) = \frac{G_U(x,y)}{h(x)h(y)}, \qquad x,y \in U.$$

Then, $G_U^{(h)}$ is a symmetric Green function for U with respect to $\mathcal{H}_{W,h}$ and $\{G_U^{(h)}\}_{U:P\text{-domain} \subset W}$ is a consistent system. Hence, $(W, \mathcal{H}_{W,h})$ is self-adjoint.

4-4. Fundamental properties of Green potentials (cf. [16],[6])

<u>Lemma 4.5.</u> Let U be a P-domain and let G_U be a Green function for U. If μ is a non-negative measure on U such that Supp μ is compact and contained in U, then G_U^μ defined by

$$G_U^\mu(x) = \int_U G_U(x,y)\, d\mu(y), \quad x \in U$$

is a potential on U and harmonic on $U \setminus \mathrm{Supp}\, \mu$.

<u>Proof.</u> It is easy to see that $G_U^\mu \in \mathcal{U}^+(U)$. Let V be any relatively compact open set such that $\overline{V} \subset U \setminus \mathrm{Supp}\, \mu$. Then, $x \mapsto G_U(x,y)$ is harmonic on V and uniformly bounded for $y \in \mathrm{Supp}\, \mu$. It then follows that G_U^μ is harmonic on V. Thus, G_U^μ is harmonic on $U \setminus \mathrm{Supp}\, \mu$. Then, by Remark 2.1 , $G_U^\mu \in \mathcal{S}^+(U)$. Next, let W be a relatively compact open set such that Supp $\mu \subset W$ and $\overline{W} \subset U$. Let $y_0 \in W$. By the continuity of $G_U(x,y)$ on $U \times U \setminus \{(x,x) \mid x \in U\}$, we see that there is $M > 0$ such that

$$G_U(x,y) \leq M\, G_U(x,y_0)$$

for all $x \in \partial W$ and $y \in \mathrm{Supp}\, \mu$. By Proposition 2.5 , this inequality holds for all $x \in U \setminus W$ and $y \in \mathrm{Supp}\, \mu$. Hence

$$G_U^\mu(x) \leq M\mu(U) G_U(x,y_0), \qquad x \in U \setminus \overline{W}.$$

Since $\mu(U) < +\infty$, Corollary 2.4 implies that $G_U^\mu \in \mathcal{P}(U)$.

Proposition 4.6. Let U be a P-domain and let G_U be a Green function on U. If μ is a non-negative measure on U such that $G_U^{\mu}(x_o) < +\infty$ for some $x_o \in U$, then $G_U^{\mu} \in \mathcal{P}(U)$ and is harmonic on $U \setminus \mathrm{Supp}\,\mu$.

Proof. By Remark 2.1, $G_U^{\mu} \in \mathcal{S}^+(U)$. Let $\{W_n\}$ be an exhaustion of U such that $x_o \in W_1$ and put

$$p_n = G_U^{\mu|W_n} \quad , \qquad h_n = G_U^{\mu} - p_n = G_U^{\mu|U\setminus W_n} .$$

Then, $p_n \in \mathcal{P}(U)$ by the above lemma and we see that $h_n|W_n \in \mathcal{H}(W_n)$, $\{h_n\}$ is monotone decreasing, $h_n \geq 0$ and $h_n(x_o) \to 0$ $(n \to \infty)$. Hence, $h_n \to 0$ on U by Axiom 3 and Lemma 1.1. If $u \in \mathcal{H}(U)$ and $u \leq G_U^{\mu}$, then $u - h_n \leq p_n$. Since $u - h_n \in -\mathcal{S}(U)$, we have $u - h_n \leq 0$. Hence, letting $n \to \infty$, we see that $u \leq 0$, that is, G_U^{μ} is a potential. By the above lemma, p_n is harmonic on $U \setminus \mathrm{Supp}\,\mu$, and hence $G_U^{\mu} = \lim_{n\to\infty} p_n$ is harmonic on $U \setminus \mathrm{Supp}\,\mu$ by Axiom 3.

Proposition 4.7. Let U be a P-domain such that there is a bounded strictly positive superharmonic function on U. Let G_U be a symmetric Green function on U. If μ is a non-negative measure on U such that $\mu(U) < +\infty$, then $G_U^{\mu} \in \mathcal{P}(U)$.

Proof. Fix $x_o \in U$ and choose a relatively compact open set V such that $x_o \in V$ and $\overline{V} \subset U$. By Lemma 4.5 , $G_U^{\mu|V} \in \mathcal{P}(U)$. Let $s_o \in \mathcal{S}^+(U)$ be bounded and strictly positive on U. Then,

$$G_U(x_o, y) \leq \alpha s_o(y) \qquad \text{for all } y \in \partial V$$

for some α. Since G_U is symmetric, $y \mapsto G_U(x_o, y)$ is a potential. Hence, by Proposition 2.5 ,

$$G_U(x_o, y) \leq \alpha s_o(y) \qquad \text{for all } y \in U \setminus V.$$

Hence,

$$G_U^{\mu|U\setminus V}(x_o) \leq \alpha\mu(U\setminus V)(\sup_U s_o) < +\infty.$$

Thus, $G_U^{\mu|U\setminus V} \in \mathcal{P}(U)$ by the previous proposition. Hence $G_U^{\mu} \in \mathcal{P}(U)$.

Proposition 4.8. Let U be a P-domain such that there is $s_o \in \mathcal{S}(U)$ which is bounded on U and for which $\inf_U s_o > 0$. Let G_U be a symmetric Green function on U. If μ,ν are non-negative measures on U and $G_U^{\mu} \leq G_U^{\nu}$ on U, then $\mu(U) \leq \beta\nu(U)$, where

$$\beta = \frac{\sup_U s_o}{\inf_U s_o}.$$

In particular, if $1 \in \mathcal{S}(U)$, then we can take $\beta=1$.

Proof. Let W be any relatively compact open set such that $\overline{W} \subset U$. Put $p_W = R_U(s_o\chi_W)$. By Proposition 2.7 , $p_W \in \mathcal{P}(U)$. By the Theorem of Hervé, there is a non-negative measure λ_W such that $p_W = G_U^{\lambda_W}$. Obviously, $p_W \leq s_o$ on U and $p_W = s_o$ on W. Hence

$$\mu(W) \leq \frac{1}{\inf_U s_o}\int_U p_W \, d\mu = \frac{1}{\inf_U s_o}\int_U G_U^{\mu}\, d\lambda_W$$

$$\leq \frac{1}{\inf_U s_o}\int_U G_U^{\nu}\, d\lambda_W = \frac{1}{\inf_U s_o}\int_U p_W \, d\nu \leq \beta\, \nu(U).$$

Since W is arbitrary, we have $\mu(U) \leq \beta\nu(U)$.

4-5. Canonical measure representation (cf. [24])

Hereafter, we assume that $(X, \mathcal{H})$ is a self-adjoint harmonic space. A measure representation associated with a consistent system of symmetric Green functions on X will be called a canonical measure representation of $\mathcal{R}$. Canonical measure representation is uniquely determined up to a multiplicative constant.

Thus, hereafter, we shall fix a consistent system $\{G_U\}_{U:\text{P-domain}}$

<u>of symmetric Green functions on X</u> <u>and the associated canonical measure representation</u> σ. We recall that to each $f \in \mathcal{R}(U)$ and a P-domain $V \subset U$ such that $f|V \in \mathcal{S}_C(V) - \mathcal{S}_C(V)$,

$$f|V = G_V^{\sigma(f)} + u_f^V$$

with $u_f^V \in \mathcal{H}(V)$, where

$$G_V^{\sigma(f)}(x) = \int_V G_V(x,y)\, d\sigma(f)(y).$$

In general, for $\nu \in \mathcal{M}(U)$ (U:P-domain), if $G_U^{|\nu|}$ is finite everywhere, then we denote by G_U^{ν} the function $G_U^{\nu+} - G_U^{\nu-}$.

If $W \in \mathcal{O}_X$ and $h \in C(W)$ is strictly positive on W, the canonical measure representation of $\mathcal{R}_W^{(h)}$ associated with the consistent system $\{G_U^{(h)}\}_{U:\text{P-domain} \subset W}$, where $G_U^{(h)}$ are given by (4.3), will be denoted by $\sigma^{(h)}$: $\sigma^{(h)}(f) = h\sigma(fh)$ for $f \in \mathcal{R}^{(h)}(U)$, $U \subset W$. (Note that this notation $\sigma^{(h)}$ does not accord with the notation in Part I.)

<u>Lemma 4.6.</u> Let U be a P-domain and let $f \in \mathcal{R}(U)$. If μ is a non-negative measure on U such that G_U^{μ} is a potential and $\mu \leq |\sigma(f)|$, then G_U^{μ} is continuous on U.

<u>Proof.</u> For any open set $V \subset U$ such that $f|V = s_1 - s_2$ with $s_1, s_2 \in \mathcal{S}_C(V)$, $s_1 + s_2 - G_U^{\mu}$ is superharmonic on V, so that it is lower semi-continuous. It follows that G_U^{μ} is continuous on V. Since such V's cover U, G_U^{μ} is continuous on U.

<u>Proposition 4.9.</u> Let U be a P-domain and V be a relatively compact domain such that $\overline{V} \subset U$. Then, for any $f \in \mathcal{R}(U)$, $G_V^{|\sigma(f)|}$ is bounded continuous on V and

$$f|V = u + G_V^{\sigma(f)}$$

with $u \in \mathcal{H}(V)$, which is also bounded on V.

Proof. By Lemma 4.5 , $G_U^{|\sigma(f)||V}$ is a potential on U, and by Lemma 4.6, it is continuous. Hence, $G_U^{|\sigma(f)||V}$ is bounded on V. Since $G_V(x,y) \leq G_U(x,y)$, $G_V^{|\sigma(f)|} \leq G_U^{|\sigma(f)||V}$. Therefore, $G_V^{|\sigma(f)|}$ is a bounded potential on V. Again by Lemma 4.6 , it is continuous. Furthermore, $u = f|V - G_V^{\sigma(f)}$ is bounded harmonic on V.

A relatively compact domain V such that $\overline{V}$ is contained in another P-domain will be called a PC-domain. For a domain U, let

$$\mathcal{P}_{BC}(U) = \{p \in \mathcal{P}(U) \mid p \text{ is bounded continuous on } U\}$$

and

$$\mathcal{Q}_{BC}(U) = \mathcal{P}_{BC}(U) - \mathcal{P}_{BC}(U).$$

The above proposition shows that if $f \in \mathcal{R}(U)$ and V is a PC-domain such that $\overline{V} \subset U$, then $f|V = u + g$ with $u \in \mathcal{H}(V)$ and $g \in \mathcal{Q}_{BC}(V)$. By Proposition 4.1 , any PC-domain is resolutive, and $u = H_f^V$ in the above case.

4-6. PB-domains (cf. [24])

Hereafter, we also assume that $1 \in \mathcal{R}(X)$. Thus, $\tilde{\mathcal{R}}(U) = \mathcal{R}(U)$ for any $U \in \mathcal{O}_X$ and $\delta_{[f,g]}$ and δ_f considered with respect to the canonical representation σ are defined for $f,g \in \mathcal{R}(U)$.

A P-domain U will be called a PB-domain if $G_U^{|\sigma(1)|}$ is bounded on U. A PC-domain is a PB-domain by Proposition 4.9.
If $1 \in \mathcal{S}(U)$, i.e., $\sigma(1) \geq 0$, then $1 = u + G_U^{\sigma(1)}$ with $u \in \mathcal{H}(U)$, $u \geq 0$. Hence, $0 \leq G_U^{\sigma(1)} \leq 1$, so that U is a PB-domain. Thus, if $1 \in \mathcal{S}(X)$, then any P-domain is a PB-domain.
If U is a PB-domain, then $G_U^{\sigma(1)^+}$, $G_U^{\sigma(1)^-} \in \mathcal{P}_{BC}(U)$ by Lemma 4.6 , so that $G_U^{\sigma(1)} \in \mathcal{Q}_{BC}(U)$. Obviously,

$$1 = u_U + G_U^{\sigma(1)} \qquad \text{with } u_U \in \mathcal{H}(U). \tag{4.4}$$

Since $G_U^{\sigma(1)^+} \geq 1 - u_U \geq - u_U$, $u_U \geq 0$. Thus, $G_U^{\sigma(1)} \leq 1$. Put

$$(4.5) \qquad s_U = 1 + G_U^{\sigma(1)^-}.$$

s_U is bounded continuous superharmonic on U and $s_U \geq 1$ on U. Put

$$(4.6) \qquad \beta_U = \sup_U s_U.$$

Then, $\beta_U \geq 1$; and $\beta_U = 1$ if and only if $\sigma(1) \geq 0$ on U.
Since $G_U^{\sigma(1)^+} = s_U - u_U \leq s_U$ and $G_U^{\sigma(1)^-} = s_U - 1$, we have

$$(4.7) \qquad G_U^{\sigma(1)^+} \leq \beta_U, \; G_U^{\sigma(1)^-} \leq \beta_U - 1, \; G_U^{|\sigma(1)|} \leq 2\beta_U - 1.$$

Also, since $1 \geq G_U^{\sigma(1)} \geq - G_U^{\sigma(1)^-} \geq - (\beta_U - 1)$,

$$(4.7)' \qquad |G_U^{\sigma(1)}| \leq \max(1, \beta_U - 1).$$

Lemma 4.7. Let U be a PB-domain. If $v \in \mathcal{S}(U)$ is bounded continuous on U, then $v = u+p$ with $u \in \mathcal{H}(U)$ and $p \in \mathcal{P}_{BC}(U)$.

Proof. Let $|v| \leq M$ on U. Then $v \geq -Ms_U$, so that v has a Riesz decomposition $v = u+p$ with $u \in \mathcal{H}(U)$ and $p \in \mathcal{P}(U)$. Obviously p is continuous. Since $-\beta_U M \leq - Ms_U \leq u \leq v \leq M$,

$$|p| = |v-u| \leq (1+\beta_U)M.$$

Hence, p is bounded, so that $p \in \mathcal{P}_{BC}(U)$.

Lemma 4.8. Let U be a PB-domain and let s be a bounded continuous superharmonic function on U. Then s^2 is a difference of two bounded continuous superharmonic functions on U. In fact,

$$v = 2Ms - s^2 + M^2 G_U^{\sigma(1)^-}$$

is superharmonic on U, where $M = \sup_U |s|$.

Proof. By Theorem 3.1 ,

$$2s\sigma(s) - \sigma(s^2) - s^2\sigma(1) = \delta_s \geq 0.$$

Since $\sigma(s) \geq 0$, it follows that

$$2M\sigma(s) - \sigma(s^2) + M^2\sigma(1)^- \geq 0.$$

Hence $\sigma(v) \geq 0$, so that v is superharmonic on U. Obviously, v is bounded continuous and

$$s^2 = (2Ms + M^2 G_U^{\sigma(1)^-}) - v.$$

Corollary 4.2. Let U be a PB-domain and $s_1, s_2 \in \mathcal{S}_C(U)$ be bounded. Then

$$s_1 s_2 = u + g$$

with $u \in \mathcal{H}(U)$ and $g \in \mathcal{Q}_{BC}(U)$.

Proof. By the previous two lemmas, $s_1^2 = u_1+g_1$, $s_2^2 = u_2+g_2$ and $(s_1+s_2)^2 = u_3+g_3$ with $u_i \in \mathcal{H}(U)$ and $g_i \in \mathcal{Q}_{BC}(U)$, $i = 1,2,3$. Since $s_1 s_2 = \frac{1}{2}\{(s_1+s_2)^2 - s_1^2 - s_2^2\}$, we obtain the assertion of the corollary.

Corollary 4.3. Let U be a PB-domain, s be a bounded continuous superharmonic function and $p \in \mathcal{P}_{BC}(U)$. Then $sp \in \mathcal{Q}_{BC}(U)$.

Proof. By the previous corollary,

$$sp = u + g \qquad \text{with } u \in \mathcal{H}(U) \text{ and } g \in \mathcal{Q}_{BC}(U).$$

Let $g = p_1 - p_2$ with $p_1, p_2 \in \mathcal{P}_{BC}(U)$ and let $M = \sup_U |s|$. Since $|sp| \leq Mp$, we have $|u| \leq p_1+p_2+Mp$, which implies $u = 0$. Thus $sp = g \in \mathcal{Q}_{BC}(U)$.

Lemma 4.9. Let U be a PB-domain. If $p \in \mathcal{P}_{BC}(U)$ and $\sigma(p)(U) < +\infty$, then

$$\int_U p^2 \, d|\sigma(1)| < +\infty, \quad |\sigma(p^2)|(U) < +\infty \text{ and } \delta_p(U) < +\infty.$$

Proof. Let $M = \sup_U p$ and put

$$q = 2G_U^{p\sigma(p)} + G_U^{p^2\sigma(1)^-}.$$

Since $0 \leq p\sigma(p) \leq M\sigma(p)$ and $0 \leq p^2\sigma(1)^- \leq M^2\sigma(1)^-$, we see that $q \in \mathcal{P}_{BC}(U)$ (Lemma 4.6). Furthermore,

$$\sigma(q) - \sigma(p^2) = 2p\sigma(p) + p^2\sigma(1)^- - \sigma(p^2) \geq \delta_p \geq 0. \tag{4.8}$$

Hence, $q-p^2$ is superharmonic. Since $|q-p^2| \leq q+Mp \in \mathcal{P}(U)$, it follows that $q-p^2 \in \mathcal{P}_{BC}(U)$. Furthermore,

$$\sigma(q)(U) = 2\int_U p \, d\sigma(p) + \int_U p^2 \, d\sigma(1)^-$$

$$\leq 2M\sigma(p)(U) + M\int_U G_U^{\sigma(1)^-} \, d\sigma(p) \leq (\beta_U+1)M\sigma(p)(U) < +\infty$$

by (4.7). Since $q-p^2 \leq q$, Proposition 4.8 implies that $\sigma(q-p^2)(U) < +\infty$. Hence $\delta_p(U) < +\infty$ by (4.8) and

$$|\sigma(p^2)|(U) \leq \sigma(q)(U) + \sigma(q-p^2)(U) < +\infty.$$

Also,

$$\int_U p^2 \, d|\sigma(1)| \leq M \int_U G_U^{|\sigma(1)|} \, d\sigma(p) \leq M(2\beta_U-1)\sigma(p)(U) < +\infty$$

by (4.7).

Proposition 4.10. Let U be a PB-domain. If $g_1, g_2 \in \mathcal{Q}_{BC}(U)$ and $|\sigma(g_i)|(U) < +\infty$, $i = 1,2$, then

$$\sigma(g_1 g_2)(U) = \int_U g_1 g_2 \, d\sigma(1)$$

and

$$\delta_{[g_1,g_2]}(U) = \int_U g_1 \, d\sigma(g_2) - \int_U g_1 g_2 \, d\sigma(1).$$

<u>Proof.</u> The second equality immediately follows from the first equality and the definition of $\delta_{[g_1,g_2]}$. To prove the first equality, it is enough to consider the case $g_1 = g_2 \in \mathcal{P}_{BC}(U)$.

Thus, assuming $p \in \mathcal{P}_{BC}(U)$, $\sigma(p)(U) < +\infty$, we shall prove

(4.9) $$\sigma(p^2)(U) = \int_U p^2 \, d\sigma(1).$$

Note that by the previous lemma, $|\sigma(p^2)|(U) < +\infty$ and $\int_U p^2 \, d|\sigma(1)| < +\infty$.

We may assume $p \not\equiv 0$. Then $p > 0$ on U. For $\alpha > 0$, put

$$p_\alpha = \min(\tfrac{p}{\alpha}, 1).$$

Then, $0 \leq p_\alpha \leq 1$ for each α and $p_\alpha \uparrow 1$ as $\alpha \downarrow 0$. Consider the function

$$g_\alpha = \min(\tfrac{p}{\alpha} + G_U^{\sigma(1)^-}, 1 + G_U^{\sigma(1)^-}) = p_\alpha + G_U^{\sigma(1)^-}.$$

Since $1 + G_U^{\sigma(1)^-} = s_U \in \mathcal{S}_C(U)$ and $\frac{p}{\alpha} + G_U^{\sigma(1)^-} \in \mathcal{P}_{BC}(U)$, we see that $g_\alpha \in \mathcal{P}_{BC}(U)$. Since $|\sigma(p^2)|(U) < +\infty$, we have

$$\sigma(p^2)(U) = \lim_{\alpha\to 0} \int_U p_\alpha \, d\sigma(p^2)$$

$$= \lim_{\alpha\to 0} \int_U g_\alpha \, d\sigma(p^2) - \int_U G_U^{\sigma(1)^-} \, d\sigma(p^2)$$

$$= \lim_{\alpha\to 0} \int_U p^2 \, d\sigma(g_\alpha) - \int_U p^2 \, d\sigma(1)^-.$$

Let $U_\alpha = \{x \in U \mid p(x) > \alpha\}$. Then U_α is an open set and $p_\alpha = 1$ on U_α. It follows that $\sigma(g_\alpha)|U_\alpha = \sigma(1)^+|U_\alpha$. Hence

$$\int_U p^2 \, d\sigma(g_\alpha) = \int_{U_\alpha} p^2 \, d\sigma(1)^+ + \int_{U\setminus U_\alpha} p^2 \, d\sigma(g_\alpha).$$

Since $U_\alpha \uparrow U$ as $\alpha \downarrow 0$,

$$\lim_{\alpha\to 0} \int_{U_\alpha} p^2 \, d\sigma(1)^+ = \int_U p^2 \, d\sigma(1)^+.$$

On the other hand,

$$0 \le \int_{U\setminus U_\alpha} p^2 \, d\sigma(g_\alpha) \le \alpha \int_{U\setminus U_\alpha} p \, d\sigma(g_\alpha)$$

$$\le \alpha \int_U p \, d\sigma(g_\alpha) = \alpha \int_U g_\alpha \, d\sigma(p)$$

$$\le \alpha \int_U (1+G_U^{\sigma(1)^-}) \, d\sigma(p) \le \alpha\beta_U \sigma(p)(U) \to 0$$

$$(\alpha \to 0).$$

Hence

$$\sigma(p^2)(U) = \int_U p^2 \, d\sigma(1)^+ - \int_U p^2 \, d\sigma(1)^- = \int_U p^2 \, d\sigma(1),$$

which is the required equality (4.9).

Corollary 4.4. Let U be a P-domain and suppose $\sigma(1) \ge 0$ on U. If $g \in \mathcal{Q}_{BC}(U)$ and $|\sigma(g)|(U) < +\infty$, then

$$\int_U g \, d\sigma(g) \ge 0.$$

Proof. $$\int_U g \, d\sigma(g) = \delta_g(U) + \int_U g^2 \, d\sigma(1) \ge 0.$$

4-7. Energy principle (cf. [24], [23])

Theorem 4.1. For any P-domain U, G_U is of positive type, i.e.,

$$2\int_U\int_U G_U(x,y)d\mu(x)d\nu(y) \le \int_U\int_U G_U(x,y)d\mu(x)d\mu(y) + \int_U\int_U G_U(x,y)d\nu(x)d\nu(y)$$

for any non-negative measures μ and ν on U.

Proof. First suppose $\sigma(1) \geq 0$ on U. Note that U is a PB-domain in this case. If either $G_U^\mu \equiv +\infty$ or $G_U^\nu \equiv +\infty$, then our assertion is trivial. Thus, suppose $p = G_U^\mu$ and $q = G_U^\nu$ are both potentials. By Proposition 2.9 , we can find sequences $\{p_n\}$, $\{q_n\}$ such that $p_n, q_n \in \mathcal{P}_{BC}(U)$, $\sigma(p_n)$ and $\sigma(q_n)$ have compact supports in U and $p_n \uparrow p$, $q_n \uparrow q$ $(n \to \infty)$. Here, boundedness of p_n, q_n follows from Proposition 2.5 and our assumption $1 \in \mathcal{S}(U)$. Put

$$f_{m,n} = p_m - q_n, \qquad m,n = 1,2,\ldots .$$

Then, $f_{m,n} \in \mathcal{Q}_{BC}(U)$ and $|\sigma(f_{m,n})|(U) < +\infty$. Hence, by Corollary 4.4,

$$\int_U f_{m,n}\, d\sigma(f_{m,n}) \geq 0,$$

i.e.,

$$\int_U p_m\, d\sigma(q_n) + \int_U q_n\, d\sigma(p_m) \leq \int_U p_m\, d\sigma(p_m) + \int_U q_n\, d\sigma(q_n).$$

Hence,

$$2\int_U p_m\, d\sigma(q_n) \leq \int_U p\, d\sigma(p_m) + \int_U q\, d\sigma(q_n)$$

$$= \int_U p_m\, d\sigma(p) + \int_U q_n\, d\sigma(q) \leq \int_U p\, d\sigma(p) + \int_U q\, d\sigma(q).$$

Letting $m \to \infty$, and then $n \to \infty$, we obtain

$$2\int_U p\, d\sigma(q) \leq \int_U p\, d\sigma(p) + \int_U q\, d\sigma(q),$$

which is the required inequality.

Next, let us consider the general case. Since U is P-domain, there is $p_o \in \mathcal{P}(U)$ such that $p_o > 0$ on U. Consider the harmonic space $(U, \mathcal{H}_{U,p_o})$. Since $\sigma^{(p_o)}(1) \geq 0$, the above result implies

$$2 \int_U \int_U G_U^{(p_o)}(x,y) p_o(x) p_o(y) \, d\mu(x) d\nu(y)$$

$$\leq \int_U \int_U G_U^{(p_o)}(x,y) p_o(x) p_o(y) \, d\mu(x) d\mu(y)$$

$$+ \int_U \int_U G_U^{(p_o)}(x,y) p_o(x) p_o(y) \, d\nu(x) d\nu(y).$$

In view of the relation $G_U^{(p_o)}(x,y) = G_U(x,y)/p_o(x)p_o(y)$, we obtain the inequality of the theorem.

For a P-domain U and a non-negative measure μ on U, let

$$I_U(\mu) = \int_U G_U^{\mu} \, d\mu = \int_U \int_U G_U(x,y) \, d\mu(x) d\mu(y).$$

We define

$$\mathfrak{M}_I(U) = \{\nu \in \mathfrak{M}(U) \mid I_U(|\nu|) < +\infty\}.$$

Corollary 4.5. Let U be a P-domain and let $\nu \in \mathfrak{M}_I(U)$. Then $G_U^{|\nu|}$ is μ-integrable for any non-negative $\mu \in \mathfrak{M}_I(U)$, so that, in particular, $\int_U G_U^{\nu} \, d\nu$ is well-defined, and

$$\int_U G_U^{\nu} \, d\nu \geq 0.$$

Proof. By the previous theorem,

$$2\int G_U^{|\nu|} \, d\mu \leq I_U(|\nu|) + I_U(\mu) < +\infty.$$

Hence, $G_U^{|\nu|}$ is μ-integrable. Applying the theorem to ν^+ and ν^-, we obtain the required inequality.

For $\nu \in \mathfrak{M}_I(U)$, we also write $I_U(\nu) = \int_U G_U^{\nu} \, d\nu$.

Corollary 4.6. Let U be a P-domain and $\nu_1,\nu_2 \in \mathfrak{M}_I(U)$.

Then

$$(\int_U G_U^{\nu_1}\, d\nu_2)^2 \leq I_U(\nu_1)\cdot I_U(\nu_2),$$

$\nu_1 + \nu_2 \in \mathfrak{M}_I(U)$ and

$$I_U(\nu_1+\nu_2)^{1/2} \leq I_U(\nu_1)^{1/2} + I_U(\nu_2)^{1/2}.$$

In fact, $\mathfrak{M}_I(U)$ is a linear space and $\nu \to I_U(\nu)^{1/2}$ defines a seminorm on $\mathfrak{M}_I(U)$.

Proof. To obtain the first inequality, we only have to apply the previous corollary to $\nu = \nu_1 - \nu_2$. Then the rest of assertions are easily shown.

Theorem 4.2. (Energy principle) Let U be a P-domain. If $\nu \in \mathfrak{M}_I(U)$ and $I_U(\nu) = 0$, then $\nu = 0$, i.e., $\nu \mapsto I_U(\nu)^{1/2}$ defines a norm on $\mathfrak{M}_I(U)$.

Proof. By the above corollary, $I_U(\nu) = 0$ implies that

$$\int_U G_U^{\tau}\, d\nu = 0 \qquad \text{for any } \tau \in \mathfrak{M}_I(U).$$

Let $f \in \mathcal{C}_0(U)$. By Theorem 2.3 , for any $\varepsilon > 0$ there are $p,q \in \mathcal{P}(U) \cap \mathcal{C}(U)$ such that Supp $\sigma(p)$, Supp $\sigma(q) \subset$ Supp f, Supp $(p-q) \subset$ Supp f and $|f - (p-q)| < \varepsilon$ on U. Let $\tau = \sigma(p) - \sigma(q)$. Since p,q are bounded on Supp f, we see that $\tau \in \mathfrak{M}_I(U)$. Hence

$$\int_U (p-q)\, d\nu = \int_U G_U^{\tau}\, d\nu = 0,$$

so that

$$|\int_U f\, d\nu| \leq \varepsilon |\nu|(\text{Supp } f).$$

Since $\varepsilon > 0$ is arbitrary, it follows that $\int_U f d\nu = 0$. Hence $\nu = 0$.

Remark 4.2. The above results in this subsection hold true even if $(X, \mathcal{H})$ is not self-adjoint, but if G_U is a symmetric Green function on U. For, by Proposition 4.4, $(U, \mathcal{H}_U)$ is self-adjoint and we can apply the above results to this space.

Remark 4.3. Energy principle implies the following principles (cf. [23; Theorem 4.1], [8], [9]): Let U be a P-domain.

(a) (Cartan's maximum principle) Let $\mu \in \mathcal{M}_I^+(U)$ and $s \in \mathcal{S}^+(U)$. If $s \geq G_U^\mu$ on Supp μ, then $s \geq G_U^\mu$ on U.

(b) (Domination principle) Let $p \in \mathcal{P}(U)$ be locally bounded on U and let $s \in \mathcal{S}^+(U)$. If $s \geq p$ on Supp $\sigma(p)$, then $s \geq p$ on U.

(c) (Continuity principle) If $s \in \mathcal{S}(U)$ and $s|\mathrm{Supp}\,\sigma(s)$ is continuous, then s is continuous on U.

4-8. Green's formula for functions in $\mathcal{Q}_{IC}(U)$ (cf. [24])

For a P-domain U, let

$$\mathcal{P}_{IC}(U) = \{p \in \mathcal{P}(U) \mid p \text{ is continuous and } \sigma(p) \in \mathcal{M}_I(U)\}$$

and

$$\mathcal{Q}_{IC}(U) = \mathcal{P}_{IC}(U) - \mathcal{P}_{IC}(U).$$

$\mathcal{P}_{IC}(U)$ is a convex cone and $\mathcal{Q}_{IC}(U)$ is a linear space.

Lemma 4.10. If $p_n, p \in \mathcal{P}_{IC}(U)$, $n = 1,2,\ldots$ and $p_n \uparrow p$, then $I_U(\sigma(p_n)-\sigma(p)) \to 0$ $(n \to \infty)$ and $I_U(\sigma(p_n)) \uparrow I_U(\sigma(p))$.

Proof. Since $\{p_n\}$ is increasing, $\{I_U(\sigma(p_n))\}$ is increasing. By Lebesgue's convergence theorem, we have

$$0 \leq I_U(\sigma(p_n) - \sigma(p)) = \int_U (p-p_n)\, d\sigma(p) - \int_U (p-p_n)\, d\sigma(p_n)$$

$$\leq \int_U (p-p_n)\, d\sigma(p) \to 0 \quad (n \to \infty).$$

Hence, we obtain the lemma.

Lemma 4.11. Let U be a P-domain and μ be a non-negative measure on U such that $G_U^\mu \leq 1$ on U. Then, for any $g \in \mathcal{Q}_{IC}(U)$, $g\mu \in \mathcal{M}_I(U)$ and

$$I_U(g\mu) \leq I_U(\sigma(g)).$$

Proof. For simplicity, let us omit the subscripts U in $G_U^\cdot$, $I_U(\cdot)$ and $\int_U \cdot$. If $\mu = 0$, then the assertion is trivial. Thus, we assume that $G^\mu > 0$ on U.

First, consider the set

$$S = \{\nu \in \mathcal{M}_I(U) \mid |G^\nu| \leq 1, \int |G^\nu|\, d\mu \leq 1\}$$

and the mapping $A : S \to \mathcal{M}(U)$ defined by

$$A\nu = G^\nu \mu \qquad \text{for } \nu \in S.$$

If $\nu \in S$, then

$$|G^{A\nu}| \leq G^{|G^\nu|\mu} \leq G^\mu \leq 1.$$

$$\int |G^{A\nu}|\, d\mu \leq \int G^{|G^\nu|\mu}\, d\mu = \int G^\mu |G^\nu|\, d\mu \leq \int |G^\nu|\, d\mu \leq 1,$$

$$I(|A\nu|) = \int G^{|G^\nu|\mu} |G^\nu|\, d\mu \leq \int G^\mu |G^\nu|\, d\mu \leq 1.$$

Hence, $A(S) \subset S$ and $\{I(A\nu) \mid \nu \in S\}$ is bounded. We also have

$$(4.10) \qquad I(A\nu) = \int G^{A\nu} G^\nu\, d\mu = \int G^\nu\, d(G^{A\nu}\mu) = \int G^\nu d(A^2\nu) \leq I(\nu)^{\frac{1}{2}} \cdot I(A^2\nu)^{\frac{1}{2}}.$$

From this, it follows that $I(A\nu) \leq I(\nu)$ for all $\nu \in S$. In fact, suppose $I(A\nu_o) > I(\nu_o)$ for some $\nu_o \in S$. If $I(\nu_o) = 0$, then $\nu_o = 0$, so that $I(A\nu_o) = 0$ which is impossible. Hence $I(\nu_o) > 0$. Let $k = I(A\nu_o)/I(\nu_o)$. By (4.10)

$$\frac{I(A^2\nu_o)}{I(A\nu_o)} \geq \frac{I(A\nu_o)}{I(\nu_o)} = k,$$

and hence $I(A^2\nu_o) \geq k^2 I(\nu_o)$. By induction, we see that

$$I(A^m\nu_o) \geq k^m I(\nu_o) \to +\infty,$$

since $k > 1$. This contradicts the boundedness of $\{I(A\nu) \mid \nu \in S\}$. Thus, we have shown

$$I(G^{\nu}\mu) \leq I(\nu) \tag{4.11}$$

for all $\nu \in S$.

Next, let $g \in \mathcal{Q}_{IC}(U)$ and Supp $\sigma(g)$ be compact in U. Since there is a bounded strictly positive potential G^{μ}, we see by Proposition 2.5 that g is bounded. Also

$$\int |g| \, d\mu \leq \int G^{\mu} \, d|\sigma(g)| \leq |\sigma(g)|(U) < +\infty.$$

Hence, for some $\alpha > 0$, $\alpha\sigma(g) \in S$, so that by (4.11)

$$\alpha^2 I(g\mu) \leq \alpha^2 I(\sigma(g)),$$

i.e.,

$$I(g\mu) \leq I(\sigma(g)). \tag{4.12}$$

Finally, let $g \in \mathcal{Q}_{IC}(U)$ be arbitrary. Let $g = p-q$ with $p,q \in \mathcal{P}_{IC}(U)$. By Proposition 2.9, we can choose $\{p_n\}$ and $\{q_n\}$ in $\mathcal{P}_{IC}(U)$ such that Supp $\sigma(p_n)$, Supp $\sigma(q_n)$ are compact in U for each n and $p_n \uparrow p$, $q_n \uparrow q$. Put $g_n = p_n - q_n$. Then, by (4.12)

$$I(g_n\mu) \leq I(\sigma(g_n)).$$

By Lemma 4.10, we see that $I(\sigma(g_n)) \to I(\sigma(g))$. On the other hand, again by (4.12),

$$I(p_n\mu) \leq I(\sigma(p_n)) \leq I(\sigma(p)) \text{ and } I(q_n\mu) \leq I(\sigma(q_n)) \leq I(\sigma(q)).$$

By monotone convergence theorem $G^{p_n\mu} \uparrow G^{p\mu}$ and $G^{q_n\mu} \uparrow G^{q\mu}$, which imply $I(p_n\mu) \uparrow I(p\mu)$ and $I(q_n\mu) \uparrow I(q\mu)$. Hence

$$I(p\mu) \leq I(\sigma(p)) < +\infty \text{ and } I(q\mu) \leq I(\sigma(q)) < +\infty,$$

that is $g\mu \in \mathcal{M}_I(U)$. Then, by Lemma 4.10 again, $I(g_n\mu) \to I(g\mu)$. Therefore, we obtain the required inequality.

Lemma 4.12. Let U be a P-domain and μ be a non-negative measure on U such that G_U^μ is bounded on U. Then

$$\int_U g^2 \, d\mu \leq (\sup_U G_U^\mu) \, I_U(\sigma(g))$$

for all $g \in \mathcal{Q}_{IC}(U)$.

Proof. Let $\beta = \sup_U G_U^\mu$. By the previous lemma, we have

$$I_U(g\mu) \leq \beta^2 \, I_U(\sigma(g)).$$

Hence

$$\int_U g^2 d\mu = \int_U g d(g\mu) \leq I_U(\sigma(g))^{\frac{1}{2}} \cdot I_U(g\mu)^{\frac{1}{2}}$$

$$\leq \beta \, I_U(\sigma(g)).$$

Theorem 4.3. Let U be a PB-domain.

(a) If $f \in \mathcal{Q}_{IC}(U)$, then $\delta_f(U) < +\infty$, $\int_U f^2 \, d|\sigma(1)| < +\infty$ and $\delta_f(U) = I_U(\sigma(f)) - \int_U f^2 \, d\sigma(1)$.

(b) For any $f_1, f_2 \in \mathcal{Q}_{IC}(U)$,

$$\delta_{[f_1,f_2]}(U) = \int_U f_1 \, d\sigma(f_2) - \int_U f_1 f_2 \, d\sigma(1).$$

Proof. (a) It is enough to prove the assertion for $f \in \mathcal{P}_{IC}(U)$. Since U is a PB-domain, the above lemma implies that

$$\int_U f^2 \, d|\sigma(1)| < +\infty.$$

Let $\{W_n\}$ be an exhaustion of U and choose $\varphi_n \in \mathcal{C}(U)$ such that $\varphi_n = 1$ on W_n, $\varphi_n = 0$ on $U \setminus W_{n+1}$ and $0 \le \varphi_n \le 1$ on U. Put

$$f_n = R_U(\varphi_n f), \qquad n = 1,2,\ldots .$$

Then, $f_n \in \mathcal{P}_{BC}(U)$, $0 \le f_n \le f$, $f_n = f$ on W_n, Supp $\sigma(f_n)$ is compact in U for each n and $f_n \uparrow f$. By Proposition 4.10,

(4.13) $$\delta_{f_n}(U) = I_U(\sigma(f_n)) - \int_U f_n^2 \, d\sigma(1).$$

Hence,

$$\delta_{f_n}(U) \le I_U(\sigma(f)) + \int_U f^2 \, d\sigma(1)^-.$$

Since $f_n = f$ on W_n, it follows that

$$\delta_f(W_n) \le I_U(\sigma(f)) + \int_U f^2 \, d\sigma(1)^- < +\infty.$$

Letting $n \to \infty$, we see that $\delta_f(U) < +\infty$. Next, let $n < m$. By Proposition 4.10 again

$$\delta_{f_n - f_m}(U) = I_U(\sigma(f_n) - \sigma(f_m)) - \int_U (f_n - f_m)^2 \, d\sigma(1).$$

By Lemma 4.10 , $I_U(\sigma(f_n) - \sigma(f_m)) \to I_U(\sigma(f_n) - \sigma(f))$ as $m \to \infty$, and by Lebesgue's convergence theorem,

$$\int_U (f_n - f_m)^2 \, d\sigma(1) \to \int_U (f_n - f)^2 \, d\sigma(1) \qquad (m \to \infty).$$

Hence

$$\delta_{f_n-f}(W_m) = \delta_{f_n-f_m}(W_m)$$

$$\leq \delta_{f_n-f_m}(U) \to I_U(\sigma(f_n) - \sigma(f)) - \int_U (f_n-f)^2\, d\sigma(1)$$

$$(m \to \infty).$$

Therefore,

$$\delta_{f_n-f}(U) \leq I_U(\sigma(f_n) - \sigma(f)) - \int_U (f_n-f)^2\, d\sigma(1).$$

Again, by Lemma 4.10 and Lebesgue's convergence theorem, the right hand side tends to zero as $n \to \infty$. Hence, $\delta_{f_n-f}(U) \to 0$ $(n \to \infty)$, so that $\delta_{f_n}(U) \to \delta_f(U)$. Thus, letting $n \to \infty$ in (4.13), we obtain the required equality.

The assertion (b) follows immediately from (a).

<u>Corollary 4.7.</u> Let U be a PB-domain. Then, for any $f \in \mathcal{Q}_{IC}(U)$

$$\delta_f(U) \leq \beta_U\, I_U(\sigma(f))$$

$$\int_U f^2\, d\sigma(1)^- \leq (\beta_U-1) I_U(\sigma(f))$$

and

$$I_U(\sigma(f)) \leq \delta_f(U) + \int_U f^2\, d|\sigma(1)| \leq (2\beta_U-1) I_U(\sigma(f)).$$

This corollary is easily seen by Theorem 4.3, (a) and (4.7).

§5. Energy-finite harmonic functions and Green's formula

5-1. Preliminary lemmas (cf. [24])

Lemma 5.1. Let U be a PB-domain and $p\in \mathcal{P}(U)\cap \mathcal{C}(U)$. For $\alpha > 0$, let

$$U(p;\alpha) = \{x\in U \mid p(x) > \alpha\}.$$

Then $U(p;\alpha)^a$ is resolutive and $\rho_x^{U(p;\alpha)}(\partial^a U(p;\alpha)\setminus U) = 0$ for all $x\in U(p;\alpha)$.

Proof. Let e_p be an Evans function for p (see Lemma 4.3). For any $\varepsilon > 0$, there is a compact set K_ε in U such that $\varepsilon e_p \geq p$ on $U\setminus K_\varepsilon$. Since $p > \alpha$ on $U(p;\alpha)$, it follows that $e_p > \alpha/\varepsilon$ on $U(p;\alpha)\setminus K_\varepsilon$. Hence

$$\lim_{x\in U(p;\alpha),x\to\xi} e_p(x) = +\infty$$

for all $\xi\in\partial^a U(p;\alpha)\setminus U$, which means that $e_p \mid U(p;\alpha)$ is an Evans function on $U(p;\alpha)$ for the set $\partial^a U(p;\alpha)\setminus U$. Hence by Proposition 4.1 , we obtain the assertion of the present lemma.

Lemma 5.2. Let U, p, $U(p;\alpha)$ be as in the previous lemma. Suppose furthermore $\sigma(p)(U) < +\infty$. Then

$$|\sigma(1)|(U(p;\alpha)) \leq \frac{2\beta_U-1}{\alpha}\,\sigma(p)(U)$$

and

$$\lim_{\alpha\to 0} \alpha\, |\sigma(1)|(U(p;\alpha)) = 0.$$

Proof.

$$|\sigma(1)|(U(p;\alpha)) \leq \frac{1}{\alpha}\int_{U(p;\alpha)} p\, d|\sigma(1)|$$

$$\leq \frac{1}{\alpha}\int_U p\, d|\sigma(1)| = \frac{1}{\alpha}\int_U G_U^{|\sigma(1)|}\, d\sigma(p) \leq \frac{2\beta_U-1}{\alpha}\,\sigma(p)(U).$$

Next, let

$$\varepsilon_o = \limsup_{\alpha\to 0} \alpha|\sigma(1)|(U(p;\alpha)).$$

Suppose $\varepsilon_o > 0$. Then we could choose by induction a sequence $\{\alpha_n\}$ such that

$$\alpha_1 > \dots > \alpha_n > \alpha_{n+1} > \dots \to 0,$$

$$\alpha_{n+1}|\sigma(1)|(U(p;\alpha_n)) < \frac{\varepsilon_o}{3}, \quad \alpha_{n+1}|\sigma(1)|(U(p;\alpha_{n+1})) > \frac{2}{3}\varepsilon_o.$$

Put $U_n = U(p;\alpha_n)$ for simplicity. Then,

$$+\infty > (2\beta_U-1)\sigma(p)(U) \geq \int_U G_U^{|\sigma(1)|}\, d\sigma(p) = \int_U p\, d|\sigma(1)|$$

$$\geq \sum_{n=1}^{\infty} \int_{U_{n+1}\setminus U_n} p\, d\sigma(1) \geq \sum_{n=1}^{\infty} \alpha_{n+1}\{|\sigma(1)|(U_{n+1}) - |\sigma(1)|(U_n)\}$$

$$\geq \sum_{n=1}^{\infty} \frac{\varepsilon_o}{3} = +\infty,$$

a contradiction. Therefore, $\varepsilon = 0$, i.e., $\lim_{\alpha\to o} \alpha|\sigma(1)|(U(p;\alpha))=0$.

<u>Lemma 5.3.</u> Let U be a PB-domain and $p\in \mathcal{P}(U)\cap \mathcal{C}(U)$. Suppose that $\sigma(p)$ has compact support in U. Let

$$U(p;\alpha) = \{x\in U \mid p(x) > \alpha\}$$

for $\alpha>0$. Then there are a non-negative measure $\mu_{p,\alpha}$ and a (signed) measure $\pi_{p,\alpha}$ on U satisfying the following conditions:

(a) Supp $\mu_{p,\alpha} \subset \partial U(p;\alpha)$, Supp $\pi_{p,\alpha} \subset \partial U(p;\alpha)$;

(b) $\mu_{p,\alpha}(U) < +\infty$, $|\pi_{p,\alpha}|(U) < +\infty$;

(c) For any bounded function Ψ on $\partial^a U(p;\alpha)$ which is continuous on $\partial U(p;\alpha)\cap U$,

(5.1) $$\int \Psi \, d\mu_{p,\alpha} = \int_{U(p;\alpha)} H_\Psi^{U(p;\alpha)^a} \, d\sigma(p).$$

(5.2) $$\int \Psi \, d\pi_{p,\alpha} = \int_{U(p;\alpha)} H_\Psi^{U(p;\alpha)^a} \, d\sigma(1);$$

(d) If we put

$$w_{p,\alpha} = \frac{1}{\alpha} G_U^{\mu_{p,\alpha}} - G_U^{\pi_{p,\alpha}} + G_U^{\sigma(1)|U(p;\alpha)},$$

then $w_{p,\alpha} = 1$ on $U(p;\alpha)$ for $0 < \alpha < \inf_{\sigma(p)} p$ and

$|w_{p,\alpha}| \leq 2\beta_U$ on U for any $\alpha > 0$.

Proof. If $\Psi \in \mathcal{C}(\partial^a U(p;\alpha))$, then

$$|H_\Psi^{U(p;\alpha)^a}| \leq (\sup |\Psi|) s_U \leq \beta_U \sup |\Psi|$$

and $H_\Psi^{U(p;\alpha)^a} \geq 0$ for $\Psi \geq 0$. Hence, the mapping

$$\Psi \mapsto \int_{U(p;\alpha)} H_\Psi^{U(p;\alpha)^a} \, d\sigma(p)$$

is a non-negative linear functional on $\mathcal{C}(\partial^a U(p;\alpha))$, so that there is a non-negative measure $\mu_{p,\alpha}$ on $\partial^a U(p;\alpha)$ satisfying (5.1) for all $\Psi \in \mathcal{C}(\partial^a U(p;\alpha))$. By Lemma 5.1 , we see that

$$\mu_{p,\alpha}(\partial^a U(p;\alpha) \setminus U) = 0.$$

Hence, we may regard $\mu_{p,\alpha}$ as a measure on U. Then Supp $\mu_{p,\alpha} \subset \partial U(p;\alpha)$ and $\mu_{p,\alpha}(U) < +\infty$. Also, we easily have (5.1) for bounded functions Ψ on $\partial^a U(p;\alpha)$ which are continuous on $\partial U(p;\alpha) \cap U$. Similarly, by virtue of Lemma 5.2 , the mapping

$$\Psi \mapsto \int_{U(p;\alpha)} H_\Psi^{U(p;\alpha)^a} \, d\sigma(1)$$

is a bounded linear functional on $\mathcal{C}(\partial^a U(p;\alpha))$, so that it defines a signed measure $\pi_{p,\alpha}$ on $\partial^a U(p;\alpha)$. Again, by Lemma 5.1, $\pi_{p,\alpha}$ can be regarded as a measure on U and it satisfies (a), (b) and (c) of the lemma. To show (d), put

$$q_\alpha = G_U^{\sigma(p)}|U(p;\alpha) \quad \text{and} \quad r_\alpha = G_U^{\sigma(1)}|U(p;\alpha).$$

These are bounded continuous functions on U. We shall use the same notations for the bounded functions on $\partial^a U(p;\alpha)$ which is zero on $\partial^a U(p;\alpha)\setminus U$ and is equal to q_α and r_α on $\partial U(p;\alpha)$, respectively. For $x\in U(p;\alpha)$, let

$$\Psi_x(\xi) = \begin{cases} G_U(x,\xi) & , \ \xi\in\partial U(p;\alpha)\cap U \\ 0 & , \ \xi\in\partial^a U(p;\alpha)\setminus U. \end{cases}$$

Then,

$$G_U^{\mu_{p,\alpha}}(x) = \int_U G_U(x,\xi)\, d\mu_{p,\alpha}(\xi)$$

$$= \int_{\partial U(p;\alpha)\setminus U} \Psi_x \, d\mu_{p,\alpha} = \int_{U(p;\alpha)} H_{\Psi_x}^{U(p;\alpha)^a} \, d\sigma(p)$$

$$= \int_{U(p;\alpha)} H_{\Psi_y}^{U(p;\alpha)^a}(x)\, d\sigma(p)(y) \qquad \text{(by Propsition 4.2)}$$

$$= \int_{\partial U(p;\alpha)} \{\int_{\partial U(p;\alpha)\cap U} G_U(\xi,y)\, d\rho_x^{U(p;\alpha)}(\xi)\} d\sigma(p)(y)$$

$$= \int_{\partial U(p;\alpha)\cap U} q_\alpha \, d\rho_x^{U(p;\alpha)} = H_{q_\alpha}^{U(p;\alpha)^a}(x),$$

i.e.

(5.3) $$G_U^{\mu_{p,\alpha}} = H_{q_\alpha}^{U(p;\alpha)^\alpha} \qquad \text{on} \quad U(p;\alpha).$$

Similarly, we have

(5.4) $$G_U^{\pi_{p,\alpha}} = H_{r_\alpha}^{U(p;\alpha)^a} \qquad \text{on } U(p;\alpha).$$

On the other hand, $1-r_\alpha$ is harmonic on $U(p;\alpha)$. Hence, for any $\varepsilon > 0$, $1-r_\alpha+\varepsilon e_p \in \overline{\mathcal{U}}_{1-r_\alpha}^{U(p;\alpha)^a}$ and $1-r_\alpha-\varepsilon e_p \in \underline{\mathcal{U}}_{1-r_\alpha}^{U(p;\alpha)^a}$,

where e_p is an Evans function for p (see the proof of Lemma 5.1).

It follows that

$$(5.5) \qquad 1 - r_\alpha = H_1^{U(p;\alpha)^a} - H_{r_\alpha}^{U(p;\alpha)^a}.$$

If $0 < \alpha < \inf_{\sigma(p)} p$, then Supp $\sigma(p) \subset U(p;\alpha)$, so that $q_\alpha = p$. Hence, $q_\alpha = \alpha$ on $\partial U(p;\alpha) \cap U$, so that $H_{q_\alpha}^{U(p;\alpha)^a} = \alpha H_1^{U(p;\alpha)^a}$.

Therefore, in view of (5.3), (5.4) and (5.5), we see that $w_{p,\alpha} = 1$ on $U(p;\alpha)$ in this case. Next, for any $\alpha > 0$ and for any $x \in U$, we have

$$G_U^{\mu_{p,\alpha}}(x) = \int_U G_U(x,\xi)\, d\mu_{p,\alpha}(\xi)$$

$$= \int_{U(p;\alpha)} \{\int_{\partial U(p;\alpha)} G_U(x,\xi)\, d\rho_y^{U(p;\alpha)}(\xi)\}\, d\sigma(p)(y)$$

$$\leq \int_{U(p;\alpha)} G_U(x,y)\, d\sigma(p)(y) = p(x).$$

Hence, if $x \in U \setminus U(p;\alpha)$, then $0 \leq G_U^{\mu_{p,\alpha}}(x) \leq \alpha$. Similarly, we see that

$$- G_U^{\sigma(1)^+}(x) \leq - G_U^{\pi_{p,\alpha}}(x) \leq G_U^{\sigma(1)^-}(x) \qquad \text{for } x \in U.$$

Obviously,

$$- G_U^{\sigma(1)^-}(x) \leq G_U^{\sigma(1)|U(p;\alpha)}(x) \leq G_U^{\sigma(1)^+}(x) \text{ for } x \in U.$$

Hence,

$$|w_{p,\alpha}(x)| \leq 1 + G_U^{|\sigma(1)|}(x) \leq 2\beta_U \quad \text{for } x \in U \setminus U(p;\alpha).$$

Therefore, $|w_{p,\alpha}| \leq 2\beta_U$ on U for any $\alpha > 0$.

<u>Lemma 5.4.</u> Let U be a PB-domain and V be a relatively compact open set such that $\overline{V} \subset U$. Then there exists a (signed) measure $\lambda = \lambda_V$ on U satisfying the following conditions:

(a) Supp $\lambda \subset \overline{V}$, (b) $G_U^{\lambda} = 1$ on V, $G_U^{\lambda} \geq 0$ on U,

(c) $G_U^{\lambda^-} \leq \beta_U - 1$ and $G_U^{\lambda^+} \leq \beta_U$ on U. (In particular, if $\sigma(1) \geq 0$ on U, then $\lambda \geq 0$.)

<u>Proof.</u> Put $p = R_U(s_U \chi_V)$ and $q = R_U((s_U-1)\chi_V)$, where χ_V is the characteristic function of V. By Proposition 2.7, these are potentials on U, $0 \leq q \leq p \leq s_U$ and $q \leq s_U - 1$ on U, and $p-q = 1$ on V. Let $\lambda = \sigma(p) - \sigma(q)$. By Proposition 2.6, Supp $\lambda \subset \overline{V}$. Furthermore, $G_U^{\lambda} = p-q \geq 0$ on U, $= 1$ on V, $G_U^{\lambda^-} \leq q \leq s_U - 1 \leq \beta_U - 1$ and $G_U^{\lambda^+} \leq p \leq s_U \leq \beta_U$.

<u>5-2. Bounded energy-finite harmonic functions and preliminary Green's formula</u> (cf. [24])

For $f \in \mathcal{R}(U)$ $(U \in \mathcal{O}_X)$,

$$E_U[f] = \delta_f(U) + \int_U f^2 \, d|\sigma(1)|$$

will be called the <u>energy</u> of f on U. By Theorem 4.3 , if U is a PB-domain, then any $f \in \mathcal{Q}_{IC}(U)$ has finite energy. In this subsection, we establish Green's formula for $f \in \mathcal{Q}_{BC}(U)$ with $|\sigma(f)|(U) < +\infty$ and a harmonic function u which belongs to

$$\mathcal{H}_{BE}(U) = \{u \in \mathcal{H}(U) \mid E_U(u) < +\infty, \text{ u is bounded on } U\}.$$

<u>Lemma 5.5.</u> Let U be a PB-domain. If $u \in \mathcal{H}_{BE}(U)$, $f \in \mathcal{Q}_{IC}(U)$ and $|\sigma(f)|(U) < +\infty$, then $|\sigma(uf)|(U) < +\infty$.

<u>Proof.</u> Since

$$\delta_{[u,f]} = \frac{1}{2}\{u\sigma(f) - \sigma(uf) - uf\,\sigma(1)\},$$

$$\sigma(uf) = 2\delta_{[u,f]} + u\sigma(f) - uf\,\sigma(1).$$

Hence,

$$|\sigma(uf)| \leq 2|\delta_{[u,f]}| + |u||\sigma(f)| + |uf||\sigma(1)|.$$

Therefore,

$$|\sigma(uf)|(U) \leq 2\delta_u(U)^{1/2}\,\delta_f(U)^{1/2} + (\int u^2\, d|\sigma(1)|)^{1/2}\,(\int f^2\, d|\sigma(1)|)^{1/2} + (\sup_U |u|)\,|\sigma(f)|(U) < +\infty.$$

<u>Lemma 5.6.</u> Let U be a PB-domain, $u \in \mathcal{H}_{BE}(U)$, $f \in \mathcal{Q}_{BC}(U)$ and $|\sigma(f)|(U) < +\infty$. Let $\{W_n\}$ be an exhaustion of U and put $f_n = G_U^{\sigma(f)|W_n}$. Then

$$\sigma(uf_n)(U) \to \sigma(uf)(U) \qquad (n \to \infty).$$

<u>Proof.</u> Obviously, $f, f_n \in \mathcal{Q}_{IC}(U)$ and $|\sigma(f_n)|(U) < +\infty$. Hence, by the previous lemma, $|\sigma(uf)|(U) < +\infty$ and $|\sigma(uf_n)|(U) < +\infty$ for each n. Put $M = \sup_U |u|$. Let V denote arbitrary relatively compact open set such that $\overline{V} \subset U$ and let λ_V be the measure given in Lemma 5.4. Since uf_n, $uf \in \mathcal{Q}_{BC}(U)$ by Corollary 4.3, we have

$$|\sigma(uf_n)(U) - \sigma(uf)(U)|$$

$$= \lim_{V\uparrow U} |\int_U G_U^{\lambda_V}\, d\sigma(uf_n) - \int_U G_U^{\lambda_V}\, d\sigma(uf)|$$

$$= \lim_{V\uparrow U} |\int_U (uf_n - uf)\,d\lambda_V|$$

$$\leq M \liminf_{V\uparrow U} \int_U |f_n - f|\,d|\lambda_V|$$

$$= M \liminf_{V\uparrow U} \int_U |G_U^{\sigma(f)|U\setminus W_n}|\; d|\lambda_V|$$

$$\leq M \liminf_{V\uparrow U} \int_{U\setminus W_n} G_U^{|\lambda_V|}\; d|\sigma(f)| \leq M(2\beta_U - 1)|\sigma(f)|(U\setminus W_n)$$

for each n. Hence, $\sigma(uf_n)(U) \to \sigma(uf)(U)$ as $n \to \infty$.

<u>Proposition 5.1.</u> Let U be a PB-domain. If $u \in \mathcal{H}_{BE}(U)$, $f \in \mathcal{Q}_{BC}(U)$ and $|\sigma(f)|(U) < +\infty$, then

$$\sigma(uf)(U) = \int_U u \, d\sigma(f) + \int_U uf \, d\sigma(1).$$

<u>Proof.</u> It is enough to prove the case where $f = p \in \mathcal{P}_{BC}(U)$.

First, assume that $\sigma(p)$ has compact support in U. For $\alpha > 0$, let $w_{p,\alpha}$ be the function given in Lemma 5.3. Then, by (d) of Lemma 5.3 and Corollary 4.3 , we have

$$\sigma(up)(U) = \lim_{\alpha \to 0} \int w_{p,\alpha} \, d\sigma(up)$$

$$= \lim_{\alpha \to 0} \{\frac{1}{\alpha} \int_U up \, d\mu_{p,\alpha} - \int_U up \, d\pi_{p,\alpha} + \int_{U(p;\alpha)} up \, d\sigma(1)\}.$$

Since $p = \alpha$ on $\operatorname{Supp} \mu_{p,\alpha}$ and on $\operatorname{Supp} \pi_{p,\alpha}$, using Lemma 5.3 ,(c), we have

$$\frac{1}{\alpha} \int_U up \, d\mu_{p,\alpha} = \int_U u \, d\mu_{p,\alpha} = \int_{U(p;\alpha)} H_u^{U(p;\alpha)^a} \, d\sigma(p)$$

and

$$\int_U up \, d\pi_{p,\alpha} = \alpha \int_U u \, d\pi_{p,\alpha} = \alpha \int_{U(p;\alpha)} H_u^{U(p;\alpha)^a} \, d\sigma(1).$$

Since $u+\varepsilon e_p \in \overline{\mathcal{U}}_u^{U(p;\alpha)^a}$ and $u-\varepsilon e_p \in \underline{\mathcal{U}}_u^{U(p;\alpha)^a}$ for any $\varepsilon > 0$, where e_p is an Evans function for p (see the proof of Lemma 5.1), we see that $H_u^{U(p;\alpha)^a} = u$. Hence

$$\frac{1}{\alpha} \int_U up \, d\mu_{p,\alpha} = \int_{U(p;\alpha)} u \, d\sigma(p) \to \int_U u \, d\sigma(p) \quad (\alpha \to 0)$$

and

$$|\int_U up\, d\pi_{p,\alpha}| = \alpha|\int_{U(p;\alpha)} u\, d\sigma(1)|$$

$$\leq (\sup_U |u|)\alpha|\sigma(1)|(U(p;\alpha)) \to 0 \quad (\alpha \to 0)$$

by Lemma 5.2. On the other hand, $\int_U p^2\, d|\sigma(1)| < +\infty$ by Lemma 4.9 and $\int_U u^2\, d|\sigma(1)| < +\infty$ by definition. Hence, $\int_U |up|\, d|\sigma(1)| < +\infty$, so that

$$\int_{U(p;\alpha)} up\, d\sigma(1) \to \int_U up\, d\sigma(1) \qquad (\alpha \to 0).$$

Therefore, we obtain the required equality in the case where Supp $\sigma(p)$ is compact in U.

In the general case, let $\{W_n\}$ be an exhaustion of U and put $p_n = G_U^{\sigma(p)|W_n}$. Then $\sigma(p_n) = \sigma(p)|W_n$ and $p_n \uparrow p$. By Lebesgue's convergence theorem, we have

$$\int_U u\, d\sigma(p_n) + \int_U up_n\, d\sigma(1) \to \int_U u\, d\sigma(p) + \int_U up\, d\sigma(1) \quad (n \to \infty).$$

On the other hand, by Lemma 5.6 , $\sigma(up_n)(U) \to \sigma(up)(U)$. Since the required equality holds for each $f = p_n$, so does for $f = p$.

Corollary 5.1. Let U be a PB-domain, $u \in \mathcal{H}_{BE}(U)$, $f \in \mathcal{Q}_{BC}(U)$ and $|\sigma(f)|(U) < +\infty$. Then $|\delta_{[u,p]}|(U) < +\infty$ and

$$(5.6) \qquad \delta_{[u,f]}(U) + \int_U uf\, d\sigma(1) = 0.$$

Proof. This is simply a reformulation of the previous proposition.

5-3. Green's formula on PB-domains

We now extend the formula (5.6) to functions $f \in \mathcal{Q}_{IC}(U)$ and u in

$$\mathcal{H}_E(U) = \{u \in \mathcal{H}(U) \mid E_U[u] < +\infty\}.$$

Lemma 5.7. Let U be an open set and V be a PC-domain such that $\bar{V} \subset U$. If $f \in \mathcal{R}(U)$, then

$$u = f|V - G_V^{\sigma(f)} \in \mathcal{H}_{BE}(V).$$

Proof. By Proposition 4.9 , $p = G_V^{\sigma(f)} \in \mathcal{Q}_{BC}(V)$ and u is bounded harmonic on V. Since $|\sigma(f)|(V) < +\infty$, $E_V[p] < +\infty$ by Lemma 4.9 (or Theorem 4.3). Obviously, $E_V[f] < +\infty$. Hence $E_V[u] < +\infty$, i.e., $u \in \mathcal{H}_{BE}(V)$.

Lemma 5.8. Let U be a PB-domain and let $\{W_n\}$ be an exhaustion of U such that each W_n is a domain. Let $f \in \mathcal{Q}_{IC}(U)$ and put $f_n = G_{W_n}^{\sigma(f)}$. Then

$$\delta_{f-f_n}(W_n) \to 0 \qquad \text{and} \quad \int_{W_n} (f-f_n)^2 \, d|\sigma(1)| \to 0 \quad (n \to \infty).$$

Proof. It is enough to consider the case $f \in \mathcal{P}_{IC}(U)$, i.e., $\sigma(f) \geq 0$. By Theorem 4.3, $\int_U f^2 \, d|\sigma(1)| < +\infty$. Since $G_{W_n} \uparrow G_U$ (cf. the proof of Proposition 4.4), we see that $f_n \uparrow f$. Hence, Lebesgue's convergence theorem implies

$$\int_{W_n} (f-f_n)^2 \, d|\sigma(1)| \to 0 \qquad (n\to\infty).$$

Next, let $u_n = f|W_n - f_n$. Then $u_n \in \mathcal{H}_{BE}(W_n)$ by the previous lemma. Hence, by Corollary 5.1,

$$\delta_{[u_n,f_n]}(W_n) + \int_{W_n} u_n f_n \, d\sigma(1) = 0,$$

or

$$\delta_{[f-f_n,f_n]}(W_n) + \int_{W_n} (f-f_n)f_n \, d\sigma(1) = 0.$$

On the other hand, by Theorem 4.3,

$$\delta_{f_n}(W_n) = I_{W_n}(\sigma(f)) - \int_{W_n} f_n^2 \, d\,\sigma(1),$$

$$\delta_f(U) = I_U(\sigma(f)) - \int_U f^2 \, d\sigma(1).$$

Hence

$$\begin{aligned}
\delta_{f-f_n}(W_n) &= \delta_f(W_n) - \delta_{f_n}(W_n) - 2\delta_{[f-f_n,f_n]}(W_n) \\
&\leq \delta_f(U) - \delta_{f_n}(W_n) + 2\int_{W_n}(f-f_n)f_n \, d\sigma(1) \\
&= I_U(\sigma(f)) - I_{W_n}(\sigma(f)) - \int_U f^2 \, d\sigma(1) + \int_{W_n} f_n^2 \, d\sigma(1) \\
&\qquad + 2\int_{W_n}(f-f_n)\, f_n \, d\sigma(1) \\
&= I_U(\sigma(f)) - I_{W_n}(\sigma(f)) - \int_{W_n}(f-f_n)^2 \, d\sigma(1) \\
&\qquad - \int_{U\setminus W_n} f^2 \, d\sigma(1).
\end{aligned}$$

Since $G_{W_n} \uparrow G_U$, we see that $I_{W_n}(\sigma(f)) \uparrow I_U(\sigma(f)) < +\infty$. Thus we conclude that $\delta_{f-f_n}(W_n) \to 0$ $(n \to \infty)$.

<u>Theorem 5.1.</u> Let U be a PB-domain. If $u \in \mathcal{H}_E(U)$ and $f \in \mathcal{Q}_{IC}(U)$, then

$$|\delta_{[u,f]}|(U) < +\infty, \quad \int_U |uf| \, d|\sigma(1)| < +\infty \text{ and}$$

$$\delta_{[u,f]}(U) + \int_U uf \, d\sigma(1) = 0.$$

<u>Proof.</u> Since $\delta_u(U) < +\infty$ by definition and $\delta_f(U) < +\infty$ by Theorem 4.3, $|\delta_{[u,f]}|(U) < +\infty$ by Proposition 3.3. Also, $\int_U u^2 \, d|\sigma(1)| < +\infty$ by definition and $\int_U f^2 \, d|\sigma(1)| < +\infty$ by Theorem 4.3.

Hence $\int_U |uf|\, d|\sigma(1)| < +\infty$. Now, let $\{W_n\}$ be an exhaustion of U such that each W_n is a domain. Let $f_n = G_{W_n}^{\sigma(f)}$. Since $u|W_n \in \mathcal{H}_{BE}(W_n)$, $f_n \in \mathcal{Q}_{BC}(W_n)$ and $|\sigma(f_n)|(W_n) = |\sigma(f)|(W_n) < +\infty$, Corollary 5.1 implies that

$$\delta_{[u,f_n]}(W_n) + \int_{W_n} uf_n\, d\sigma(1) = 0.$$

By the previous lemma, $\delta_{f-f_n}(W_n) \to 0$. Hence, by Proposition 3.3, we see that $\delta_{[u,f_n]}(W_n) \to \delta_{[u,f]}(U)$ $(n \to \infty)$. Also, $\int_{W_n} (f-f_n)^2\, d|\sigma(1)| \to 0$ by the previous lemma, which implies that $\int_{W_n} uf_n\, d\sigma(1) \to \int_U uf\, d\sigma(1)$ $(n \to \infty)$. Hence, we obtain the theorem.

Combining this theorem with Theorem 4.3 , we obtain the following Green's formula for PB-domains:

<u>Theorem 5.2.</u> Let U be a PB-domain. If $f \in \mathcal{H}_E(U) + \mathcal{Q}_{IC}(U)$ and $g \in \mathcal{Q}_{IC}(U)$, then $|\delta_{[f,g]}|(U) < +\infty$, $\int_U |fg|\, d|\sigma(1)| < +\infty$ and

$$\delta_{[f,g]}(U) + \int_U fg\, d\sigma(1) = \int_U g\, d\sigma(f).$$

We remark here that the space $\mathcal{H}_E(U) + \mathcal{Q}_{IC}(U)$ can be characterized as follows:

<u>Proposition 5.2.</u> If U is a PB-domain, then

$$\mathcal{H}_E(U) + \mathcal{Q}_{IC}(U) = \{f \in \mathcal{R}(U) \mid E_U[f] < +\infty \text{ and } \sigma(f) \in \mathcal{M}_I(U)\}.$$

<u>Proof.</u> We have already seen that if $f \in \mathcal{H}_E(U) + \mathcal{Q}_{IC}(U)$, then $E_U[f] < +\infty$ and $\sigma(f) \in \mathcal{M}_I(U)$. Conversely, suppose $f \in \mathcal{R}(U)$, $E_U[f] < +\infty$ and $\sigma(f) \in \mathcal{M}_I(U)$. Let $g = G_U^{\sigma(f)}$. Then $g \in \mathcal{Q}_{IC}(U)$ and $u = f - g \in \mathcal{H}(U)$. Since $E_U[g] < +\infty$ by Theorem 4.3 , we see that $E_U[u] < +\infty$, i.e., $u \in \mathcal{H}_E(U)$.

Remark 5.1. In general, $\mathcal{Q}_{IC}(U)$ does not coincide with the space $\{f \in \mathcal{Q}_C(U) \mid E_U[f] < +\infty\}$, where $\mathcal{Q}_C(U) = \mathcal{P}_C(U) - \mathcal{P}_C(U)$, $\mathcal{P}_C(U) = \mathcal{P}(U) \cap \mathcal{C}(U)$. This fact can be seen even in the classical case. However, for a PB-domain U, we have

$$\mathcal{P}_{IC}(U) = \{p \in \mathcal{P}_C(U) \mid E_U[p] < +\infty\}$$

(cf. Proposition 6.5).

5-4. Green's formula for general open sets

Theorem 5.3. Let U be an open set in X. If $f, g \in \mathcal{R}(U)$ and Supp g is compact in U, then

$$\delta_{[f,g]}(U) + \int_U fg \, d\sigma(1) = \int_U g \, d\sigma(f).$$

Proof. Note that Supp $\delta_{[f,g]} \subset$ Supp g, so that $\delta_{[f,g]}(U)$ is well-defined. Let $\{U_j\}_{j=1}^n$ be a finite covering of Supp g such that each U_j is a PC-domain and $\overline{U}_j \subset U$. By Proposition 2.19 , we can find $h_1, \ldots, h_n \in \mathcal{R}(U)$ such that $h_j \geq 0$ on U, Supp h_j is compact and contained in U_j for each j and $\Sigma_{j=1}^n h_j = 1$ on Supp g. For each j, we see that $h_j g \in \mathcal{Q}_{IC}(U_j)$. On the other hand,

$$f|U_j \in \mathcal{H}_E(U_j) + \mathcal{Q}_{IC}(U_j)$$

by Lemma 5.7. Hence, by Theorem 5.2,

$$\delta_{[f,h_j g]}(U_j) + \int_{U_j} fh_j g \, d\sigma(1) = \int_{U_j} h_j g \, d\sigma(f)$$

for each j. Since $h_j g = 0$ outside U_j,

$$\delta_{[f,h_j g]}(U) + \int_U fh_j g \, d\sigma(1) = \int_U h_j g \, d\sigma(f)$$

for each j. Adding up in j and remarking that $(\Sigma_{j=1}^n h_j)g = g$, we obtain the required formula.

Corollary 5.2. Let U be any open set in X and let $u \in \mathcal{R}(U)$. Then, $u \in \mathcal{H}(U)$ if and only if

$$\delta_{[u,g]}(U) + \int_U ug\, d\sigma(1) = 0 \tag{5.7}$$

for all $g \in \mathcal{R}(U) \cap \mathcal{C}_0(U)$.

Proof. If $u \in \mathcal{H}(U)$, then $\sigma(u) = 0$, so that (5.7) holds by the above theorem. Conversely, suppose (5.7) holds for any $g \in \mathcal{R}(U) \cap \mathcal{C}_0(U)$. Then, by the above theorem,

$$\int_U g\, d\sigma(u) = 0 \qquad \text{for all } g \in \mathcal{R}(U) \cap \mathcal{C}_0(U).$$

Let V be any P-set contained in U and let $f \in \mathcal{C}_0(V)$. By Theorem 2.3 , for any $\varepsilon > 0$ we can find $g \in \mathcal{R}(U) \cap \mathcal{C}_0(U)$ such that Supp $g \subset V$ and $|f-g| < \varepsilon$ on V. It follows that $\int_V f\, d\sigma(u) = 0$ for all $f \in \mathcal{C}_0(V)$, so that $\sigma(u) = 0$ on V. Since such V's cover U, $\sigma(u) = 0$ on U. Therefore, $u \in \mathcal{H}(U)$.

As an application of this corollary, we have

Theorem 5.4. Let U be a domain in X and $f \in \mathcal{R}(U)$. If $\delta_f = 0$ on U, then $f = $ const. on U.

Proof. First, assume that $\sigma(1) = 0$ on U. Then, since $\delta_{[f,g]} = 0$ for any $g \in \mathcal{R}(U)$ by Proposition 3.3,(c), the above corollary implies that $f \in \mathcal{H}(U)$. Then

$$0 = \delta_f = -\frac{1}{2}\sigma(f^2),$$

so that $f^2 \in \mathcal{H}(U)$. Hence for any $x_0 \in U$, $(f - f(x_0))^2 \in \mathcal{H}(U)$. By Lemma 1.1 , we conclude that $f - f(x_0) \equiv 0$ on U, i.e., $f = $ const.

Next, consider the general case. Let V be any subdomain of U on which there exists $h \in \mathcal{H}(V)$ such that $h > 0$ on V. Consider the self-adjoint harmonic space $(V, \mathcal{H}_{V,h})$ and the canonical measure representation $\sigma^{(h)}$: $\sigma^{(h)}(g) = h\sigma(gh)$ for $g \in \mathcal{R}^{(h)}(U) = \mathcal{R}(U)$ $(U \subset V)$. Let $\delta^{(h)}$ be the associated gradient measure. Then

$$\delta_f^{(h)} = h^2 \delta_f = 0 \qquad \text{on } V.$$

Since $\sigma^{(h)}(1) = 0$, it follows from the above result that f = const. on V. Since such V's cover U and since U is connected, it follows that f = const. on U.

<u>Remark 5.2.</u> It is not known whether Theorem 5.4 remains valid for any (i.e., not necessarily self-adjoint) Brelot's harmonic space.

PART III SPACES OF DIRICHLET -FINITE AND ENERGY-FINITE FUNCTIONS ON SELF-ADJOINT HARMONIC SPACES

§6. Spaces of Dirichlet-finite and energy-finite harmonic functions

6-1. Harnack's inequality (cf. [18])

First, we establish Harnack's inequality on Brelot's harmonic space. Thus, in this subsection, let $(X, \mathcal{H})$ be a Brelot's harmonic space which may or may not be self-adjoint.

Given a domain U in X and $x_o \in U$, let

$$\mathcal{H}^+_{x_o}(U) = \{u \in \mathcal{H}(U) \mid u \geq 0 \text{ on } U,\ u(x_o) = 1\}.$$

Proposition 6.1. $\mathcal{H}^+_{x_o}(U)$ is locally uniformly bounded; thus for any compact set K in U, there is $\alpha_K > 0$ such that

$$\sup_K u \leq \alpha_K\, u(x_o)$$

for all $u \in \mathcal{H}^+(U)$.

Proof. Suppose $\mathcal{H}^+_{x_o}(U)$ is not uniformly bounded on a compact set K in U. Then, there would exist $u_n \in \mathcal{H}^+_{x_o}(U)$, $n = 1,2,\ldots$, such that

$$\sup_K u_n > n^3, \qquad n = 1,2,\ldots .$$

Put

$$u = \sum_{n=1}^{\infty} \frac{u_n}{n^2} .$$

Since $u(x_o) < +\infty$, Axiom 3 implies that $u \in \mathcal{H}(U)$. Therefore, $\sup_K u < +\infty$. On the other hand

$$\sup_K u \geq \frac{1}{n^2} \sup_K u_n > n \qquad \text{for all } n,$$

which is impossible.

Corollary 6.1. Let V be a resolutive domain and $x_o \in V$. Then for any $y \in V$ there is a non-negative bounded Borel function f_y on ∂V such that $\mu_y^V = f_y \mu_{x_o}^V$.

Proof. For any $\varphi \in \mathcal{C}_o^+(\partial V)$,

$$\int \varphi \, d\mu_y^V = H_\varphi^V(y) \le \alpha_{\{y\}} H_\varphi^V(x_o) = \alpha_{\{y\}} \int \varphi \, d\mu_{x_o}^V .$$

Hence, by Radon-Nikodym theorem, there is a Borel function f_y on ∂V such that $\mu_y^V = f_y \mu_{x_o}^V$ and $0 \le f_y \le \alpha_{\{y\}}$ on ∂V.

Proposition 6.2. Let $U \in \mathcal{O}_X$ and $\mathcal{F} \subset \mathcal{H}(U)$. If $\mathcal{F}$ is locally uniformly bounded on U, then it is a relatively compact subset of $\mathcal{C}(U)$ with respect to the locally uniform convergence topology. In particular, $\mathcal{H}_{x_o}^+(U)$ is a compact set in $\mathcal{C}(U)$, in case U is a domain.

Proof. Let $\{h_n\}$ be any sequence in $\mathcal{F}$. Let V, V' be two relatively compact resolutive domains such that $\overline{V'} \subset V$, $\overline{V} \subset U$. Fix $x_o \in V$ and for each $y \in V$ choose a non-negative bounded Borel function f_y on ∂V such that $\mu_y^V = f_y \mu_{x_o}^V$ according to the previous corollary. Since $\{h_n | \partial V\}$ is uniformly bounded, there is a subsequence $\{h_{n_j}\}$ and a bounded Borel function φ on ∂V such that $h_{n_j} | \partial V \to \varphi$ $(j \to \infty)$ in the weak* - topology in $L^\infty(\partial V; \mu_{x_o}^V) = L^1(\partial V; \mu_{x_o}^V)^*$. Since $f_y \in L^1(\partial V; \mu_{x_o}^V)$, it follows that

$$(\mu^V \varphi)(y) = \int \varphi \, f_y \, d\mu_{x_o}^V = \lim_{j \to \infty} \int h_{n_j} f_y \, d\mu_{x_o}^V = \lim_{j \to \infty} h_{n_j}(y)$$

for each $y \in V$. Hence $h_{n_j} \to h = \mu^V \varphi \in \mathcal{H}(V)$ pointwise. For each m, set

$$\overline{u}_m = \sup_{j \ge m} h_{n_j} \quad \text{and} \quad \underline{u}_m = \inf_{j \ge m} h_{n_j} .$$

Then, $\overline{u}_m$ is bounded lower semicontinuous and $\underline{u}_m$ is bounded upper semicontinuous on V. Since $\overline{u}_m \downarrow h$ and $\underline{u}_m \uparrow h$ as $m \to \infty$, $\mu^{V'}\overline{u}_m \downarrow h$ and $\mu^{V'}\underline{u}_m \uparrow h$ on V'. By virtue of Dini's lemma, these convergences are locally uniform on V'. Since $\mu^{V'}\overline{u}_m \leq h_{n_m} \leq \mu^{V'}\overline{u}_m$, it follows that $h_{n_j} \to h$ locally uniformly on V'. Since we can cover U by a countable number of such V''s, we can choose by the diagonal method a subsequence of $\{h_n\}$ which converges to a harmonic function on U locally uniformly. Hence we obtain the proposition.

Theorem 6.1. (Harnack's inequality) Let U be a domain and K be a compact set in U. Then there is $\alpha_K \geq 1$ such that

$$\sup_K u \leq \alpha_K \inf_K u$$

for all $u \in \mathcal{H}^+(U)$.

Proof. Let $x_o \in K$. By virtue of Proposition 6.1 , it is enough to show that

$$\inf_{u \in \mathcal{H}^+_{x_o}(U)} \{\inf_K u\} > 0.$$

Suppose the contrary. Then, we would find $u_n \in \mathcal{H}^+_{x_o}(U)$ and $x_n \in K$ such that $u_n(x_n) \to 0$ $(n \to \infty)$. By the previous proposition and the compactness of K, we can choose a subsequence $\{u_{n_j}\}$ of $\{u_n\}$ and a subsequence $\{x_{n_j}\}$ of $\{x_n\}$ such that $u_{n_j} \to u \in \mathcal{H}(U)$ locally uniformly on U and $x_{n_j} \to x^* \in K$ $(j \to \infty)$. Then $u(x^*) = 0$. Obviously, $u \geq 0$ on U and $u(x_o) = 1$. This is a contradiction in view of Lemma 1.1.

6-2. Lattice structures (cf. [22], [24])

From now on, we shall always assume that $(X, \mathcal{H})$ is a self-adjoint harmonic space such that $1 \in \mathcal{R}(X)$ and $\{G_U\}$ is a fixed consistent system of symmetric Green functions, σ is the associated canonical measure representation and δ denotes the gradient measure defined in terms of σ.

Given $U \in \mathcal{O}_X$, we consider the following spaces of harmonic functions:

$$\mathcal{H}_D(U) = \{u \in \mathcal{H}(U) \mid \delta_u(U) < +\infty\}$$

$$\mathcal{H}_{D'}(U) = \{u \in \mathcal{H}(U) \mid \delta_u(U) + \int_U u^2 \, d\sigma(1)^- < +\infty\}$$

$$\mathcal{H}_E(U) = \{u \in \mathcal{H}(U) \mid \delta_u(U) + \int_U u^2 \, d|\sigma(1)| < +\infty\}.$$

These are linear subspaces of $\mathcal{H}(U)$. Note that if $\sigma(1) \geq 0$ on U, i.e., if $1 \in \mathcal{S}(U)$, then $\mathcal{H}_{D'}(U) = \mathcal{H}_D(U)$. Let

$$\|u\|_{D,U} = \delta_u(U)^{1/2}$$

$$\|u\|_{D',U} = \{\delta_u(U) + \int_U u^2 \, d\sigma(1)^-\}^{1/2}$$

$$\|u\|_{E,U} = \{\delta_u(U) + \int_U u^2 \, d|\sigma(1)|\}^{1/2}.$$

These are semi-norms on $\mathcal{H}_D(U)$, $\mathcal{H}_{D'}(U)$ and $\mathcal{H}_E(U)$, respectively. They are norms if and only if $\sigma(1)|U' \not\equiv 0$ for any component U' of U (cf. Theorem 5.4).

Lemma 6.1. Let U be a PB-domain. Then

$$I_U(\sigma(|u|)) \leq (2\beta_U - 1) \|u\|^2_{D',U}$$

for any $u \in \mathcal{H}_{D'}(U)$.

Proof. Let V be any relatively compact domain such that $\overline{V} \subset U$. Then $u|V \in \mathcal{H}_{BE}(V)$. Since $|u|$ is bounded on V and V is a PB-domain, the least harmonic majorant v of $|u|$ on V exists and $p = v - |u|$ is a bounded potential. Since $\sigma(p)(V) = -\sigma(|u|)(V) < +\infty$, we see that $p \in \mathcal{P}_{IC}(V)$, so that $E_V[p] < +\infty$.

Since $E_V[|u|] = E_V[u] < +\infty$, it follows that $v \in \mathcal{H}_{BE}(V)$. Hence, by Theorem 5.1,

$$\delta_{[v,p]}(V) + \int_V vp \, d\sigma(1) = 0.$$

Thus, using Theorem 4.3 , we have

$$\begin{aligned} I_V(\sigma(|u|)) &= \delta_p(V) + \int_V p^2 \, d\sigma(1) \\ &= - \delta_{[|u|,p]}(V) - \int_V |u| p \, d\sigma(1) \\ &\leq \delta_u(V)^{1/2} \cdot \delta_p(V)^{1/2} + (\int_V u^2 \, d\sigma(1)^-)^{1/2} \\ &\qquad \cdot (\int_V p^2 \, d\sigma(1)^-)^{1/2} . \\ &\leq \|u\|_{D',U} \|p\|_{D',V}. \end{aligned}$$

On the other hand, by Corollary 4.7,

$$\|p\|^2_{D',V} \leq (2\beta_V - 1) I_V(\sigma(p)) \leq (2\beta_U - 1) I_U(\sigma(|u|)).$$

Hence

$$I_V(\sigma(|u|)) \leq (2\beta_U - 1) \|u\|^2_{D',U}.$$

Letting $V \uparrow U$, we obtain the required inequality.

Given $U \in \mathcal{O}_X$ and $u, v \in \mathcal{H}(U)$, if the least harmonic majorant of $\max(u,v)$ (resp. the greatest harmonic minorant of $\min(u,v)$) exists, then we denote it by $u \underset{U}{\vee} v$ (resp. $u \underset{U}{\wedge} v$).

<u>Theorem 6.2.</u> If U is a PB-domain, then $\dot{\mathcal{H}}_{D'}(U)$ and $\mathcal{H}_E(U)$ are vector lattices with respect to the natural order, i.e., closed under operations $\underset{U}{\vee}$ and $\underset{U}{\wedge}$. Furthermore,

$$\|u \underset{U}{\vee}(-u)\|_{D',U} \leq \{1 + 3(\beta_U - 1)\}\|u\|_{D',U} \quad \text{for } u \in \mathcal{H}_{D'}(U)$$

and

$$\|u \underset{U}{\vee}(-u)\|_{E,U} \leq \{1 + 3(\beta_U - 1)\}\|u\|_{E,U} \quad \text{for } u \in \mathcal{H}_E(U).$$

(Note that if $\sigma(1) \geq 0$ on U, then $D'=D$ and $1 + 3(\beta_U-1)=1$.)

Proof. Let $u \in \mathcal{H}_{D'}(U)$ and $\nu = -\sigma(|u|)$. Then $\nu \geq 0$ and $I_U(\nu) < +\infty$ by the above lemma. Put $p = G_U^{\nu}$ and $v = |u|+p$. Then $v = u \underset{U}{\vee}(-u)$. Since $p \in \mathcal{P}_{IC}(U)$, we see that $v \in \mathcal{H}_{D'}(U)$; and $v \in \mathcal{H}_E(U)$ if $u \in \mathcal{H}_E(U)$. Thus, $\mathcal{H}_{D'}(U)$ and $\mathcal{H}_E(U)$ are vector lattices.

Let $\{W_n\}$ be an exhaustion of U, $p_n = G_{W_n}^{\nu}$ and $u_n = p|W_n - p_n$. Then, $u_n \in \mathcal{H}_E(W_n)$, $v = |u|+u_n+p_n$ on W_n and $v-u_n \geq |u|$ on W_n. Since $v|W_n - u_n \in \mathcal{H}_E(W_n)$ and $p_n \in \mathcal{P}_{IC}(W_n)$, we have by Theorems 4.3 and 5.1,

$$\delta_{p_n}(W_n) + \int_{W_n} p_n^2 \, d\sigma(1) = I_{W_n}(\nu)$$

and

$$\delta_{[v-u_n,p_n]}(W_n) + \int_{W_n} (v-u_n)p_n \, d\sigma(1) = 0.$$

Hence, remarking that $\delta_{|u|} = \delta_u$ and $v-u_n = |u|+p_n$, we have

$$I_U(\nu) = \delta_u(W_n) - \delta_{v-u_n}(W_n) + 2\delta_{[v-u_n,p_n]}(W_n)$$

$$+ \int_{W_n} u^2 \, d\sigma(1) - \int_{W_n} (v-u_n)^2 \, d\sigma(1) +$$

$$+ 2\int_{W_n} (v-u_n)p_n \, d\sigma(1)$$

$$= \delta_u(W_n) + \int_{W_n} u^2 \, d\sigma(1) - \delta_{v-u_n}(W_n) - \int_{W_n} (v-u_n)^2 \, d\sigma(1).$$

Therefore,

$$\|v-u_n\|^2_{D',W_n} = \|u\|^2_{D',W_n} + 2\int_{W_n} \{(v-u_n)^2-u^2\}\, d\sigma(1)^- - $$
$$- \int_{W_n} \{(v-u_n)^2-u^2\}\, d\sigma(1)^+ - I_{W_n}(\nu)$$
$$\le \|u\|^2_{D',U} + 2\int_{W_n} \{(v-u_n)^2-u^2\}\, d\sigma(1)^- - I_{W_n}(\nu)$$

and

$$\|v-u_n\|^2_{E,W_n} \le \|u\|^2_{E,U} + 2\int_{W_n} \{(v-u_n)^2-u^2\}\, d\sigma(1)^- - I_{W_n}(\nu)$$

if $u \in \mathcal{H}_E(U)$. By Lemma 5.8, $\|u_n\|_{E,W_n} \to 0$. Hence

(6.1) $$\|v\|^2 \le \|u\|^2 + 2\int_U (v^2-u^2)\, d\sigma(1)^- - I_U(\nu),$$

where $\|\cdot\| = \|\cdot\|_{D',U}$ if $u \in \mathcal{H}_{D'}(U)$ and $\|\cdot\| = \|\cdot\|_{E,U}$ if $u \in \mathcal{H}_E(U)$.

If $\sigma(1) \ge 0$ on U, then (6.1) immediately implies the required inequalities. Let $\sigma(1)^- \not\equiv 0$ on U. Since $v^2-u^2 \le ku^2+(1+k^{-1})p^2$ for any $k > 0$,

(6.2) $$2\int_U (v^2-u^2)\, d\sigma(1)^- \le 2k\int_U u^2\, d\sigma(1)^- + 2(1+\tfrac{1}{k})\int_U p^2\, d\sigma(1)^-.$$

By Corollary 4.7, $\int_U p^2\, d\sigma(1)^- \le (\beta_U-1)I_U(\nu)$. Hence (6.1) and (6.2) imply

$$\|v\|^2 \le (1+2k)\|u\|^2 + \{2(1+\tfrac{1}{k})(\beta_U-1) - 1\}I_U(\nu)$$

for any $k > 0$. Using Lemma 6.1 and letting $k = 2(\beta_U-1)$, we have

$$\|v\|^2 \le [1 + 4(\beta_U-1) + 2(\beta_U-1)(2\beta_U-1)]\|u\|^2$$
$$\le \{1 + 3(\beta_U-1)\}^2\|u\|^2,$$

i.e., the required inequalities.

Corollary 6.2. Let U be a P-domain and suppose $\sigma(1) \geq 0$ on U. Then $\mathcal{H}_D(U)$ is a vector lattice with respect to the natural order and

$$\|u \underset{U}{\vee} (-u)\|_{D,U} \leq \|u\|_{D,U}.$$

Corollary 6.3. Let U be a PB-domain. Then $\mathcal{H}_{BE}(U)$ and $\mathcal{H}_{BD'}(U) = \{u \in \mathcal{H}_{D'}(U) \mid u \text{ bounded}\}$ are vector lattices with respect to the natural order.

Proof. If $u \in \mathcal{H}(U)$ is bounded, then $u \underset{U}{\vee} (-u)$ is bounded since U is a PB-domain.

Open question: In case 1 is not superharmonic on a PB-domain U, is $\mathcal{H}_D(U)$ or $\mathcal{H}_{BD}(U) = \{u \in \mathcal{H}_D(U) \mid u\text{:bounded}\}$ a vector lattice? Note that if Supp $\sigma(1)^-$ is compact in U, then $\mathcal{H}_{D'}(U) = \mathcal{H}_D(U)$, so that $\mathcal{H}_D(U)$ is a vector lattice; if $\sigma(1)^-(U) < +\infty$, then $\mathcal{H}_{BD'}(U) = \mathcal{H}_{BD}(U)$, so that $\mathcal{H}_{BD}(U)$ is a vector lattice.

6-3. Boundedness (cf. [22], [24])

Lemma 6.2. Let U be a P-domain and suppose $\sigma(1) \geq 0$ on U. Let V be a non-empty relatively compact open set such that $\overline{V} \subset U$ and let λ_V be the non-negative measure given in Lemma 5.4. Then, for any $u \in \mathcal{H}_E(U)$,

$$\inf_{x \in V} \min\{(u \underset{U}{\vee} 0)(x),\ [(-u) \underset{U}{\vee} 0](x)\} \leq \left\{\frac{-\sigma(u^2)(U)}{4\lambda_V(U)}\right\}^{1/2}.$$

(Here, note that $\sigma(u^2)(U) \leq 0$, since $\sigma(1) \geq 0$.)

<u>Proof.</u> Let $\mu = -\sigma(u^2)$ and $\nu = \sigma(\min(u,0))$. These are non-negative measures on U. For simplicity, let us omit U in the notation $\underset{U}{\vee}$ and $\underset{U}{\wedge}$. Obviously

(6.3) $$\min(u,0) = u\wedge 0 + G_U^\nu .$$

Since $(u\vee 0) + (u\wedge 0) = u = \max(u,0) + \min(u,0)$, we also have

(6.4) $$\max(u,0) = u\vee 0 - G_U^\nu .$$

Hence, $|u| = u\vee(-u) - 2G_U^\nu$. Since $\mu(U) = 2\delta_u(U) + \int_U u^2\, d\sigma(1) < +\infty$, G_U^μ is a potential on U by Proposition 4.7. Hence

$$u^2 = h - G_U^\mu \qquad \text{with } h\in \mathcal{H}(U).$$

Then, $h \geq 0$. For any regular domain W such that $\overline{W} \subset U$,

$$\mu^W h^{1/2} \leq (\mu^W h)^{1/2} (\mu^W 1)^{1/2} \leq h^{1/2} \text{ on } W.$$

Hence $h^{1/2} \in \mathcal{S}(U)$. Since $h^{1/2} \geq |u|$, it follows that

$$h^{1/2} \geq u\vee(-u).$$

Therefore,

$$0 \leq [u\vee(-u)]^2 - u^2 \leq h - u^2 = G_U^\mu .$$

Hence,

$$(G_U^\nu)^2 = \tfrac{1}{4}(u\vee(-u) - |u|)^2 \leq \tfrac{1}{4}\{[u\vee(-u)]^2 - u^2\} \leq \tfrac{1}{4}\, G_U^\mu .$$

Thus,

$$\mu(U) \geq \int_U G_U^{\lambda_V}\, d\mu = \int_U G_U^\mu\, d\lambda_V$$

$$\geq 4 \int_{\overline{V}} (G_U^\nu)^2\, d\lambda_V \geq 4\{\inf_{x\in\overline{V}} G_U^\nu(x)\}^2\, \lambda_V(U),$$

so that

$$\inf_{x \in \overline{V}} G_U^{\nu}(x) \le \left\{\frac{\mu(U)}{4\lambda_V(U)}\right\}^{1/2}. \tag{6.5}$$

On the other hand, by (6.3) and (6.4) we have

$$0 = \min\{\max(u,0), -\min(u,0)\}$$

$$= \min\{u\vee 0 - G_U^{\nu}, (-u)\vee 0 - G_U^{\nu}\}$$

$$= \min\{u\vee 0, (-u)\vee 0\} - G_U^{\nu},$$

i.e., $G_U^{\nu} = \min\{u\vee 0, (-u)\vee 0\}$. Hence (6.5) is the required inequality.

Theorem 6.3. Let U be a P-domain and suppose $\sigma(1) \ge 0$ on U. Let $x_o \in U$ be fixed and put

$$\mathcal{H}_E^1(U) = \begin{cases} \{u \in \mathcal{H}_E(U) \mid \|u\|_{E,U}^2 + u(x_o)^2 \le 1\}, & \text{if } \sigma(1) = 0 \\ \{u \in \mathcal{H}_E(U) \mid \|u\|_{E,U}^2 \le 1\}, & \text{if } \sigma(1) \ne 0. \end{cases}$$

Then, $\mathcal{H}_E^1(U)$ is a locally uniformly bounded family of functions.

Proof. We omit U in the notation $\underset{U}{V}$. Since

$$|u| \le \max\{u\vee 0, (-u)\vee 0\},$$

it is enough to show that $\{u\vee 0 \mid u \in \mathcal{H}_E^1(U)\}$ is locally uniformly bounded. By virtue of Proposition 6.1 , we only have to show that

$$\{(u\vee 0)(x_o) \mid u \in \mathcal{H}_E^1(U)\}$$

is bounded. Supposing the contrary, we could choose $u_n \in \mathcal{H}_E^1(U)$ such that

$$(u_n\vee 0)(x_o) \ge n, \qquad n = 1,2,\dots .$$

<u>The case $\sigma(1) = 0$ on U:</u> In this case, $|u_n(x_o)| \leq 1$ for all n. Hence

$$[(-u_n)\vee 0](x_o) = (u_n \vee 0)(x_o) - u_n(x_o) \geq n-1, \quad n = 1,2,\ldots .$$

Let V be any relatively compact open set such that $x_o \in V$ and $\overline{V} \subset U$. Let $\alpha = \alpha_{\overline{V}}$ in the notation in Theorem 6.1. Then

$$\inf_{x\in V} (u_n \vee 0)(x) \geq \frac{n}{\alpha} \text{ and } \inf_{x\in V} [(-u_n)\vee 0](x) \geq \frac{n-1}{\alpha},$$

so that

$$\inf_{x\in V} \min\{(u_n \vee 0)(x), [(-u_n)\vee 0](x)\} \geq \frac{n-1}{\alpha}, \; n = 1,2,\ldots .$$

Hence, by the above lemma

$$\frac{-\sigma(u_n^2)(U)}{4\lambda_V(U)} \geq \frac{(n-1)^2}{\alpha^2}, \qquad n = 1,2,\ldots .$$

This is impossible, because $-\sigma(u_n^2)(U) = 2\delta_{u_n}(U) = \|u_n\|^2_{E,U} \leq 2$ for all n.

<u>The case $\sigma(1) \geq 0$ but $\sigma(1) \not\equiv 0$ on U:</u> Let

$$v_n = \frac{u_n \vee (-u_n)}{[u_n \vee (-u_n)](x_o)}, \quad n = 1,2,\ldots .$$

Then $v_n \in \mathcal{H}^+_{x_o}(U)$ for all n. Hence, by Proposition 6.2 , we can find a subsequence $\{v_{n_j}\}$ which converges to $v \in \mathcal{H}^+_{x_o}(U)$ locally uniformly on U. By Theorem 6.2 , we have

$$\|v_n\|^2_{E,U} \leq \frac{1}{n^2} \|u_n \vee (-u_n)\|^2_{E,U} \leq \frac{1}{n^2} \|u_n\|^2_{E,U} \leq \frac{1}{n^2}$$

for all n. Hence, in particular, $\int_U v_n^2 \, d\sigma(1) \leq 1/n^2$, n = 1,2,... .

On the other hand, for any compact set K in U, $v_n \geq \alpha_K$ on K for all n with $\alpha_K > 0$, which contradicts $\int_K v_n^2 \, d\sigma(1) \leq 1/n^2$ for all n when $\sigma(1)(K) > 0$.

Thus the theorem is proved.

Theorem 6.4. Let U be a PB-domain such that $\sigma(1) \not\equiv 0$ on U. Then

$$\mathcal{H}^1_{D'}(U) = \{u \in \mathcal{H}_{D'}(U) \mid \|u\|_{D',U} \leq 1\}$$

is locally uniformly bounded on U.

Proof. (I) The case $\sigma(1) \geq 0$ on U: Let V be any relatively compact open set such that $\overline{V} \subset U$ and $\sigma(1) \mid V \not\equiv 0$. Choose a relatively compact domain V' such that $\overline{V} \subset V'$ and $\overline{V'} \subset U$, and put $\alpha = \inf_V G^{\sigma(1)}_{V'}$. Then $\alpha > 0$. We shall show that if $u \in \mathcal{H}^1_{D'}(U)$, then

(6.6) $$\|u\|_{E,V} \leq \left(\frac{2}{\alpha}\right)^{1/2}.$$

Then, in view of the previous theorem, we see that $\mathcal{H}^1_{D'}(U)$ is locally uniformly bounded on V, and since V can be chosen to contain any compact set in U, we conclude that $\mathcal{H}^1_{D'}(U)$ is locally uniformly bounded on U.

To prove (6.6), let $u \in \mathcal{H}^1_{D'}(U)$. Since $\sigma(1) \geq 0$, $\sigma(u^2) \leq 0$. For simplicity, let $\mu = -\sigma(u^2)$. By Lemma 5.7 , $h = u^2 + G^{\mu}_{V'}$ belongs to $\mathcal{H}_{BE}(V')$. If $u = 0$ on V', then (6.6) is trivial; so assume that $u \not\equiv 0$ on V'. Then $h > 0$. By Lemma 2.1, $u^2h^{-1} \in -\mathcal{S}(V')$. Since $u^2h^{-1} \leq 1$, we have $u^2h^{-1} \leq 1 - G^{\sigma(1)}_{V'}$, i.e., $u^2 \leq h - h\, G^{\sigma(1)}_{V'}$. Hence

(6.7) $$G^{\mu}_{V'} \geq h\, G^{\sigma(1)}_{V'}.$$

By Corollary 4.3, $hG_{V'}^{\sigma(1)} \in \mathcal{Q}_{BC}(V')$, so that $hG_{V'}^{\sigma(1)} = G_{V'}^{\nu^+} - G_{V'}^{\nu^-}$, where $\nu = \sigma(hG_{V'}^{\sigma(1)})$. Then, by (6.7)

$$G_{V'}^{\mu+\nu^-} \geq G_{V'}^{\nu^+}.$$

Hence, by Proposition 4.8 (note that $1 \in \mathcal{S}(V')$), $\mu(V') + \nu^-(V') \geq \nu^+(V')$, i.e.,

$$\sigma(hG_{V'}^{\sigma(1)})(V') \leq \mu(V'). \tag{6.8}$$

On the other hand, by Proposition 5.1,

$$\sigma(hG_{V'}^{\sigma(1)})(V') = \int_{V'} h \, d\sigma(1) + \int_{V'} hG_{V'}^{\sigma(1)} \, d\sigma(1) \geq \int_{V'} h \, d\sigma(1).$$

Hence, by (6.8)

$$\mu(V') \geq \int_{V'} h \, d\sigma(1).$$

Thus,

$$\begin{aligned} \|u\|_{D',V'}^2 &= \|u\|_{D,V'}^2 = \delta_u(V') \\ &= \tfrac{1}{2}\{\mu(V') - \int_{V'} u^2 \, d\sigma(1)\} \\ &\geq \tfrac{1}{2} \int_{V'} (h-u^2) \, d\sigma(1) \\ &= \tfrac{1}{2} \int_{V'} G_{V'}^{\mu} \, d\sigma(1) = \tfrac{1}{2} \int_{V'} G_{V'}^{\sigma(1)} \, d\mu \geq \tfrac{\alpha}{2}\mu(V). \end{aligned}$$

Therefore,

$$\begin{aligned} \|u\|_{E,V}^2 &= \delta_u(V) + \int_V u^2 \, d\sigma(1) \\ &= \tfrac{1}{2}\{\mu(V) + \int_V u^2 \, d\sigma(1)\} \\ &= \mu(V) - \delta_u(V) \leq \mu(V) \leq \tfrac{2}{\alpha}\|u\|_{D',V'}^2 \leq \tfrac{2}{\alpha}, \end{aligned}$$

and we have shown (6.6).

(II) The case $\sigma(1)^- \not\equiv 0$: The proof in this case can be carried out in the same way as the proof of Theorem 6.3 in the case $\sigma(1) \geq 0$, $\sigma(1) \not\equiv 0$, replacing $\sigma(1)$ by $\sigma(1)^-$ and E by D'.

<u>Corollary 6.4.</u> If U is a PB-domain and $\sigma(1) \not\equiv 0$ on U, then

$$\mathcal{H}_E^1(U) = \{u \in \mathcal{H}_E(U) \mid \|u\|_{E,U} \leq 1\}$$

is locally uniformly bounded on U.

<u>Corollary 6.5.</u> If U is a P-domain, $\sigma(1) \geq 0$ and $\sigma(1) \not\equiv 0$ on U (or U is a PB-domain, $\sigma(1) \not\equiv 0$ and Supp $\sigma(1)^-$ is compact in U), then

$$\mathcal{H}_D^1(U) = \{u \in \mathcal{H}_D(U) \mid \|u\|_{D,U} \leq 1\}$$

is locally uniformly bounded on U.

<u>Corollary 6.6.</u> Let U be a PB-domain and suppose $\sigma(1) \not\equiv 0$ on U. If $u_n \in \mathcal{H}_{D'}(U)$, $n = 1,2,\ldots$, and $\|u_n\|_{D',U} \to 0$ $(n \to \infty)$, then $u_n \to 0$ and $u_n \overset{V}{_U} (-u_n) \to 0$, both locally uniformly on U.

<u>Corollary 6.7.</u> Let U be a P-domain such that $\sigma(1) = 0$ on U. If $u_n \in \mathcal{H}_D(U)$, $n = 1,2,\ldots$, and $\|u_n\|_{D,U} \to 0$ $(n \to \infty)$, then there is a sequence $\{c_n\}$ of constants such that $u_n + c_n \to 0$ $(n \to \infty)$ locally uniformly on U.

<u>Corollary 6.8.</u> Let U be a PB-domain and let V be a relatively compact open set such that $\overline{V} \subset U$. Then there is a constant $M = M(U,V) > 0$ such that

$$\|u\|_{E,V} \leq M \|u\|_{D',U}$$

for all $u \in \mathcal{H}_{D'}(U)$.

Proof. If $\sigma(1) = 0$ on U, then there is nothing to prove ($M = 1$). Suppose $\sigma(1) \not\equiv 0$ on U. By Theorem 6.4, there is $M' = M'(U,V) > 0$ such that $|u| \leq M'$ on V for all $u \in \mathcal{H}^1_{D'}(U)$. Hence, for any $u \in \mathcal{H}^1_{D'}(U)$

$$\int_V u^2 \, d\sigma(1)^+ \leq M'^2 \sigma(1)^+(V) \|u\|^2_{D',U} ,$$

so that

$$\|u\|^2_{E,V} \leq \{1 + M'^2 \sigma(1)^+(V)\} \|u\|^2_{D',U} .$$

6-4. Completeness (cf. [22], [24])

Proposition 6.3. Let U be any open set in X and let $u_n \in \mathcal{H}(U)$, $n = 1,2,\ldots$ If $\{u_n\}$ is locally uniformly bounded and $u_n \to 0$ pointwise on U, then

$$\delta_{u_n}(K) \to 0 \qquad (n \to \infty)$$

for any compact set K in U.

Proof. By Proposition 2.17, there is $\varphi \in \mathcal{R}(U)$ such that $0 \leq \varphi \leq 1$ on U, $\varphi = 1$ on K and $\operatorname{Supp} \varphi$ is compact in U. By Theorem 5.3,

$$\delta_{[u_n, u_n \varphi^2]}(U) + \int_U u_n^2 \varphi^2 \, d\sigma(1) = 0, \qquad n = 1,2,\ldots .$$

By Theorem 3.2, $\delta_{[u_n, u_n \varphi^2]} = \varphi^2 \delta_{u_n} + 2 u_n \varphi \, \delta_{[u_n, \varphi]}$. Hence

$$\int_U \varphi^2 \, d\delta_{u_n} = -2 \int_U u_n \varphi \, d\delta_{[u_n,\varphi]} - \int_U u_n^2 \varphi^2 \, d\sigma(1).$$

By using Proposition 3.3, (a) and approximating u_n and φ by simple Borel functions, we see that

$$-2 u_n \varphi \delta_{[u_n,\varphi]} \leq \frac{1}{2} \varphi^2 \delta_{u_n} + 2 u_n^2 \delta_\varphi .$$

Hence

$$\int_U \varphi^2 d\delta_{u_n} \le \frac{1}{2}\int_U \varphi^2\, d\delta_{u_n} + 2\int_U u_n^2\, d\delta_\varphi + \int_U u_n^2\varphi^2\, d\sigma(1)^-,$$

so that

$$\delta_{u_n}(K) \le \int_U \varphi^2\, d\delta_{u_n} \le 4\int_U u_n^2\, d\delta_\varphi + 2\int_U u_n^2\varphi^2\, d\sigma(1)^-.$$

Since Supp φ, Supp δ_φ are compact in U, Lebesgue's convergence theorem implies that the right hand side of the above inequality converges to 0 as $n \to \infty$. Hence $\delta_{u_n}(K) \to 0$ $(n \to \infty)$.

<u>Proposition 6.4.</u> Let $U \in \mathcal{O}_X$, $u_n \in \mathcal{H}(U)$, $n = 1,2,\ldots$ and $u_n \to u$ locally uniformly on U. If $u_n \in \mathcal{H}_D(U)$ (resp. $\mathcal{H}_{D'}(U)$, $\mathcal{H}_E(U)$) and $\{\|u_n\|\}$ is bounded, where $\|\cdot\| = \|\cdot\|_{D,U}$ (resp. $\|\cdot\|_{D',U}$, $\|\cdot\|_{E,U}$), then $u \in \mathcal{H}_D(U)$ (resp. $\mathcal{H}_{D'}(U)$, $\mathcal{H}_E(U)$) and

$$\|u\| \le \liminf_{n\to\infty} \|u_n\|. \tag{6.9}$$

<u>Proof.</u> Let $u_n \in \mathcal{H}_D(U)$ and $u_n \to u$ locally uniformly on U. Clearly, $u \in \mathcal{H}(U)$. Applying the previous proposition to $u_n - u$, we have

$$\delta_{u_n - u}(K) \to 0 \qquad (n \to \infty)$$

for any compact set K in U, so that

$$\delta_u(K) = \lim_{n\to\infty} \delta_{u_n}(K) \le \liminf_{n\to\infty} \|u_n\|_{D,U}^2$$

for any compact set K in U. Hence, we have (6.9) for $\|.\| = \|.\|_{D,U}$ and we see that $u \in \mathcal{H}_D(U)$. If $u_n \in \mathcal{H}_{D'}(U)$ (resp. $\mathcal{H}_E(U)$), then by Fatou's lemma

$$\int_U u^2 \, d\sigma(1)^- \le \liminf_{n \to \infty} \int_U u_n^2 \, d\sigma(1)^-$$

$$(\text{resp. } \int_U u^2 \, d|\sigma(1)| \le \liminf_{n \to \infty} \int_U u_n^2 \, d|\sigma(1)|),$$

so that we also obtain (6.9) for $\|\cdot\| = \|\cdot\|_{D,U}$ (resp. $\|\cdot\|_{E,U}$) and that $u \in \mathcal{H}_{D'}(U)$ (resp. $\mathcal{H}_E(U)$).

<u>Theorem 6.5.</u> Let U be a domain in X.

(a) If $\sigma(1) = 0$ on U, then $\mathcal{H}_E(U) = \mathcal{H}_{D'}(U) = \mathcal{H}_D(U)$ is a Hilbert space with respect to the norm $\|\cdot\|$ defined by

$$\|u\|^2 = \|u\|^2_{D,U} + |u(x_o)|^2 \qquad (x_o \in U : \text{fixed}).$$

(b) If $\sigma(1) \ne 0$ on U, then $\mathcal{H}_E(U)$ is a Hilbert space with respect to the norm $\|\cdot\|_{E,U}$.

<u>Proof.</u> In the case (b) too, let $\|\cdot\| = \|\cdot\|_{E,U}$. Then, in both cases (a) and (b), $\|\cdot\|$ is a norm on $\mathcal{H}_E(U)$ and

$$(u,v) = \delta_{[u,v]}(U) + u(x_o)v(x_o) \qquad \text{in the case (a),}$$

$$(u,v) = \delta_{[u,v]}(U) + \int_U uv \, d|\sigma(1)| \qquad \text{in the case (b)}$$

is the corresponding inner product. Hence $\mathcal{H}_E(U)$ with this norm is a pre-Hilbert space.

To prove that $\mathcal{H}_E(U)$ is complete, let $\{u_n\}$ be a Cauchy sequence in $\mathcal{H}_E(U)$ with respect to this norm. Let

$$A = \{x \in U \mid \lim_{n \to \infty} u_n(x) \text{ exists}\}.$$

If V is a PB-domain in U such that either $\sigma(1) = 0$ on V and $V \cap A \ne \emptyset$ or $\sigma(1) \ne 0$ on V, then Theorem 6.3 implies that $\{u_n\}$

is locally uniformly convergent on V, and hence $V \subset A$. Furthermore, in this case $u_V = \lim_{n\to\infty} (u_n|V)$ belongs to $\mathcal{H}_E(V)$ and

$$\|u_n - u\|_{E,V} \leq \liminf_{m\to\infty} \|u_n - u_m\|_{E,V} \to 0 \qquad (n\to\infty)$$

by Proposition 6.4. If $x\in A$ and W is a PB-domain such that $x\in W \subset U$, then from the above argument we see that $W \subset A$. Hence A is open. If $x\notin A$ and W is a PB-domain such that $x\in W \subset U$, then again by the above argument we see that $W \cap A = \emptyset$. Therefore, $U \setminus A$ is also open. Since $A \neq \emptyset$ and U is connected, it follows that $A = U$, i.e., $\{u_n\}$ converges everywhere. Again the above argument shows that $\{u_n\}$ converges locally uniformly on U. Let $u = \lim_{n\to\infty} u_n$. Then $u\in \mathcal{H}(U)$ and for any PB-domain $V \subset U$, $u|V\in \mathcal{H}_E(V)$ and $\|u_n - u\|_{E,V} \to 0$ $(n \to \infty)$. Since any compact set K in U can be covered by a finite number of PB-domains in U,

$$\|u\|_{E,K} = \lim_{n\to\infty} \|u_n\|_{E,K} \leq \lim_{n\to\infty} \|u_n\|_{E,U}$$

and

$$\|u-u_n\|_{E,K} = \lim_{m\to\infty} \|u_m-u_n\|_{E,K} \leq \lim_{m\to\infty} \|u_m-u_n\|_{E,U}.$$

Hence $u\in \mathcal{H}_E(U)$ and $\|u-u_n\|_{E,U} \to 0$ $(n \to \infty)$, i.e., $\mathcal{H}_E(U)$ is complete with respect to $\|\cdot\|$.

Theorem 6.6. If U is a domain such that $\sigma(1) \neq 0$ on U, then $\mathcal{H}_{D'}(U)$ is a Hilbert space with respect to the norm $\|\cdot\|_{D',U}$.

Proof. Obviously, $\|\cdot\|_{D',U}$ is a norm on $\mathcal{H}_{D'}(U)$ and

$$(u,v) = \delta_{[u,v]}(U) + \int_U uv\, d\sigma(1)^-$$

is the corresponding inner product. If V is a PB-domain in U and W is a PC-domain such that $\overline{W} \subset V$, then by Corollary 6.8,

$$\|u\|_{E,W} \leq M \|u\|_{D',V}$$

for all $u \in \mathcal{H}_{D'}(V)$. Hence if $\{u_n\}$ is a Cauchy sequence in $\mathcal{H}_{D'}(U)$, then $\{u_n|W\}$ is a Cauchy sequence in $\mathcal{H}_E(W)$ for any W as above. Hence, by the previous theorem, $u = \lim_{n\to\infty} u_n$ exists and $u|W \in \mathcal{H}_E(W)$ for any PC-domain W such that $\overline{W} \subset U$ and $\sigma(1)|W \not\equiv 0$. Now, by arguments similar to that in the proof of the previous theorem, we see that $u_n \to u \in \mathcal{H}_{D'}(U)$ and $\|u-u_n\|_{D',U} \to 0$ $(n \to \infty)$. Hence $\mathcal{H}_{D'}(U)$ is a Hilbert space.

Open question: Is $\mathcal{H}_D(U)$ complete with respect to $\|\cdot\|_{D,U}$ for any domain U?

6-5. The space of energy-finite functions and decomposition theorems

For an open set U in X, let

$$\mathcal{R}_E(U) = \{f \in \mathcal{R}(U) \mid E_U[f] < +\infty\}.$$

$\mathcal{R}_E(U)$ is a linear space and

$$\|f\|_{E,U} = E_U[f]^{1/2} = \{\delta_f(U) + \int_U f^2\, d|\sigma(1)|\}^{1/2}$$

defines a semi-norm on $\mathcal{R}_E(U)$. If U is a domain and $\sigma(1) \not\equiv 0$ on U, then this is a norm (cf. Theorem 5.4). By Proposition 3.5, $\mathcal{R}_E(U)$ is closed under max. and min. operations and

$$\||f|\|_{E,U} = \|f\|_{E,U} \qquad \text{for } f \in \mathcal{R}_E(U).$$

If U is a PB-domain, then $\mathcal{Q}_{IC}(U) \subset \mathcal{R}_E(U)$ (Theorem 4.3). Proposition 5.2 shows that

$$\mathcal{H}_E(U) + \mathcal{Q}_{IC}(U) = \{f \in \mathcal{R}_E(U) \mid \sigma(f) \in \mathcal{M}_I(U)\} \subset \mathcal{R}_E(U)$$

for any PB-domain U.

Lemma 6.3. Let U be a PB-domain. If $f = u + g$ with $u \in \mathcal{H}_E(U)$ and $g \in \mathcal{Q}_{IC}(U)$, then

$$\|u\|_{E,U} \leq M(\beta_U)\|f\|_{E,U}$$

and

$$\|g\|_{E,U} \leq (2\beta_U - 1)\|f\|_{E,U},$$

where

$$M(t) = \begin{cases} 2t & \text{if} \quad t \geq \frac{3}{2} \\ \min(2t, (3-2t)^{-1}) & \text{if } 1 \leq t \leq \frac{3}{2}, \end{cases}$$

in particular, $M(1) = 1$.

Proof. Using Corollary 4.7 , Theorems 4.3 and 5.1 , we have

$$\|g\|^2_{E,U} \leq (2\beta_U-1)I_U(\sigma(g))$$

$$= (2\beta_U-1)\{\delta_g(U) + \int_U g^2 \, d\sigma(1)\}$$

$$= (2\beta_U-1)\{\delta_{[f,g]}(U) + \int_U fg \, d\sigma(1)\}$$

$$\leq (2\beta_U-1)[\delta_f(U)^{1/2}\cdot\delta_g(U)^{1/2} + \{\int_U f^2 \, d|\sigma(1)|\}^{1/2} \cdot\{\int_U g^2 \, d|\sigma(1)|\}^{1/2}]$$

$$\leq (2\beta_U-1)\|f\|_{E,U}\cdot\|g\|_{E,U}.$$

Hence

$$\|g\|_{E,U} \leq (2\beta_U-1)\|f\|_{E,U}.$$

It then follows that

$$\|u\|_{E,U} \leq \|f\|_{E,U} + \|g\|_{E,U} \leq 2\beta_U\|f\|_{E,U}.$$

On the other hand, again by Theorems 4.3 and 5.1,

$$\delta_u(U) + \int_U u^2 \, d\sigma(1) = \delta_f(U) + \int_U f^2 \, d\sigma(1) - I_U(\sigma(g)),$$

so that

$$\|u\|^2_{E,U} = \|f\|^2_{E,U} - 2 \int_U g(2f-g) \, d\sigma(1)^- - I_U(\sigma(g))$$

$$\leq \|f\|^2_{E,U} + 2\,\varepsilon \int_U f^2 \, d\sigma(1)^- + 2(\tfrac{1}{\varepsilon}+1) \int_U g^2 \, d\sigma(1)^- -$$

$$- I_U(\sigma(g))$$

for any $\varepsilon > 0$. Since $\int_U g^2 \, d\sigma(1)^- \leq (\beta_U - 1)\, I_U(\sigma(g))$, if $\beta_U < \frac{3}{2}$ then choosing $\varepsilon = 2(\beta_U-1)(3-2\beta_U)^{-1}$ we have

$$\|u\|^2_{E,U} \leq \{1 + 4(\beta_U-1)(3-2\beta_U)^{-1}\}\|f\|^2_{E,U}$$

$$\leq \{1 + 2(\beta_U-1)(3-2\beta_U)^{-1}\}^2\|f\|^2_{E,U}$$

$$= (3-2\beta_U)^{-2}\|f\|^2_{E,U}.$$

Hence

$$\|u\|_{E,U} \leq (3-2\beta_U)^{-1}\|f\|_{E,U} \qquad \text{if } \beta_U < \tfrac{3}{2}.$$

Remark 6.1. Let U be a PB-domain. If $f = u + g$ with $u \in \mathcal{H}^+_E(U)$ and $g \in \mathcal{P}_{IC}(U)$, then

$$\|u\|^2_{E,U} + \|g\|^2_{E,U} \leq \|f\|^2_{E,U}.$$

For, by Theorems 4.3 and 5.1, we have

$$\|f\|^2_{E,U} = \|u\|^2_{E,U} + \|g\|^2_{E,U} + 4 \int_U ug \, d\sigma(1)^- \geq \|u\|^2_{E,U} + \|g\|^2_{E,U}$$

Proposition 6.5. If U is a PB-domain, then

$$\mathcal{R}_E(U) \cap \mathcal{S}(U) = \mathcal{H}_E(U) \oplus \mathcal{P}_{IC}(U)$$

and

$$\mathcal{R}_E(U) \cap \mathcal{P}(U) = \mathcal{P}_{IC}(U).$$

Proof. The second relation is an immediate consequence of the first relation. As we remarked before, $\mathcal{P}_{IC}(U) \subset \mathcal{P}_E(U)$. Hence

$$\mathcal{H}_E(U) + \mathcal{P}_{IC}(U) \subset \mathcal{R}_E(U) \cap \mathcal{S}(U).$$

Let $f \in \mathcal{R}_E(U) \cap \mathcal{S}(U)$. Let $\{W_n\}$ be an exhaustion of U such that each W_n is a domain. Then, by Lemma 5.7,

$$f = u_n + g_n \quad \text{on } W_n \quad \text{with } u_n \in \mathcal{H}_E(W_n), \quad g_n \in \mathcal{Q}_{IC}(W_n).$$

Since $\sigma(g_n) = \sigma(f)|W_n$, we see by the above lemma,

$$I_{W_n}(\sigma(f)) \le \|g_n\|^2_{E,W_n} \le (2\beta_{W_n}-1)^2\|f\|^2_{E,W_n} \le (2\beta_U-1)^2\|f\|^2_{E,U}.$$

Thus, $\{I_{W_n}(\sigma(f))\}$ is bounded. Since $\sigma(f) \ge 0$, $I_U(\sigma(f)) =$ $= \lim_{n\to\infty} I_{W_n}(\sigma(f))$. Hence $I_U(\sigma(f)) < +\infty$, i.e., $\sigma(f) \in \mathcal{M}_I(U)$. Therefore, by Proposition 5.2, $f \in \mathcal{H}_E(U) + \mathcal{P}_{IC}(U)$. Hence

$$\mathcal{H}_E(U) + \mathcal{P}_{IC}(U) = \mathcal{R}_E(U) \cap \mathcal{S}(U).$$

Obviously, $\mathcal{H}_E(U) \cap \mathcal{P}_{IC}(U) = \{0\}$, so that the sum is direct.

Lemma 6.4. Let U be a P-domain and let $\mathcal{R}_0(U) = \mathcal{R}(U) \cap \mathcal{C}_0(U)$. Then $\mathcal{R}_0(U) \subset \mathcal{Q}_{IC}(U)$ and for any $f \in \mathcal{Q}_{IC}(U)$ there is a sequence $\{f_n\}$ in $\mathcal{R}_0(U)$ such that $f_n \to f$ locally uniformly and

$$I_U(\sigma(f_n) - \sigma(f)) \to 0 \qquad (n \to \infty).$$

Proof. If $f \in \mathcal{R}_0(U)$, then Supp $\sigma(f)$ is compact in U and $|f|$ is dominated by a potential. Hence we easily see that $f \in \mathcal{Q}_{IC}(U)$. Next, let $f \in \mathcal{P}_{IC}(U)$. Let $\{W_n\}$ be an exhaustion of U and choose $\varphi_n \in \mathcal{C}_0(U)$ such that $0 \leq \varphi_n \leq 1$ on U, $\varphi_n = 1$ on $U \setminus W_{n+1}$ and $\varphi_n = 0$ on W_n. Put $g_n = R_U(f\varphi_n)$ for each n. Then g_n is a continuous potential, $0 \leq g_n \leq f$ and $g_n = f$ on $U \setminus W_{n+1}$. Put $f_n = f - g_n$. Then $f_n \in \mathcal{R}_0(U)$. Since each g_n is harmonic on W_n, $\{g_n\}$ is monotone decreasing and $0 \leq g_n \leq f$, we see that $g_n \to 0$ locally uniformly on U. Hence $f_n \to f$ locally uniformly on U. Furthermore,

$$0 \leq I_U(\sigma(g_n)) \leq \int_U f\, d\sigma(g_n) = \int_U g_n\, d\sigma(f)$$

and $\int_U f\, d\sigma(f) < +\infty$. Hence, Lebesgue's convergence theorem implies that $I_U(\sigma(g_n)) \to 0$ $(n \to \infty)$. Hence $I_U(\sigma(f_n) - \sigma(f)) \to 0$ $(n \to \infty)$. In the general case where $f \in \mathcal{Q}_{IC}(U)$, let $f = f_1 - f_2$ with $f_1, f_2 \in \mathcal{P}_{IC}(U)$ and approximate f_1 and f_2 by functions in $\mathcal{R}_0(U)$. Then we obtain the lemma.

Proposition 6.6. Let U be a PB-domain.

(a) $\mathcal{H}_E(U) + \mathcal{R}_0(U)$ is dense in $\mathcal{R}_E(U)$ in the sense that for any $f \in \mathcal{R}_E(U)$ we can find $f_n \in \mathcal{H}_E(U) + \mathcal{R}_0(U)$ such that $\|f_n - f\|_{E,U} \to 0$ and $f_n \to f$ locally uniformly on U.

(b) $\mathcal{R}_E(U) \cap \{\mathcal{H}(U) + \mathcal{Q}_C(U)\} = \mathcal{H}_E(U) \oplus \mathcal{Q}_{EC}(U)$, where $\mathcal{Q}_{EC}(U) = \mathcal{R}_E(U) \cap \mathcal{Q}_C(U)$;

(c) $\mathcal{R}_0(U)$ is dense in $\mathcal{Q}_{EC}(U)$; in fact, given $f \in \mathcal{Q}_{EC}(U)$, there exists a sequence $\{g_n\}$ in $\mathcal{R}_0(U)$ such that $\|f - g_n\|_{E,U} \to 0$ and $g_n \to f$ locally uniformly on U.

Proof. (a) Let $f \in \mathcal{R}_E(U)$. Let $\{W_n\}$ be an exhaustion of U. Then $f|W_n = u_n + g_n$ with $u_n \in \mathcal{H}_{BE}(W_n)$ and $g_n \in \mathcal{Q}_{IC}(W_n)$ by Lemma 5.7. By the above lemma, for each n, we can choose $\varphi_n \in \mathcal{R}_0(U)$ such that Supp $\varphi_n \subset W_n$, $|\varphi_n - g_n| < 1/n$ on W_{n-1} and $I_{W_n}(\sigma(\varphi_n) - \sigma(g_n)) < 4^{-n}$. Put $a_n = I_U(\sigma(\varphi_n))$. By Lemma 6.3 , we see that $\{\|g_n\|_{E,W_n}\}$ is bounded. Since

$$\|\varphi_n - g_n\|^2_{E,W_n} \leq (2\beta_U - 1) I_{W_n}(\sigma(\varphi_n) - \sigma(g_n)) < 4^{-n}(2\beta_U - 1),$$

$\{\|\varphi_n\|_{E,U}\}$ is also bounded. Since $a_n \leq \|\varphi_n\|^2_{E,U}$, it follows that $\{a_n\}$ is bounded. Let $a_n \leq M^2$ for all n. If $m > n$, then by Theorem 4.3 and Corollary 4.7 , we have

$$\begin{aligned}
\|\varphi_n - \varphi_m\|^2_{E,U} &\leq (2\beta_U - 1) I_U(\sigma(\varphi_n) - \sigma(\varphi_m)) \\
&= (2\beta_U - 1)\{\delta_{\varphi_n - \varphi_m}(U) + \textstyle\int_U (\varphi_n - \varphi_m)^2 \, d\sigma(1)\} \\
&= (2\beta_U - 1)\{\delta_{\varphi_m}(U) - \delta_{\varphi_n}(U) - 2\delta_{[\varphi_m - \varphi_n, \varphi_n]}(U) \\
&\quad + \textstyle\int_U \varphi_m^2 \, d\sigma(1) - \int_U \varphi_n^2 \, d\sigma(1) - 2 \int_U (\varphi_m - \varphi_n)\varphi_n \\
&\qquad d\sigma(1)\} \\
&= (2\beta_U - 1)\{a_m - a_n - 2\delta_{[\varphi_m - \varphi_n, \varphi_n]}(W_n) \\
&\qquad - 2 \textstyle\int_{W_n} (\varphi_m - \varphi_n)\varphi_n \, d\sigma(1)\}.
\end{aligned}$$

On the other hand, since $g_m|W_n - g_n \in \mathcal{H}_{BE}(W_n)$, Theorem 5.1 implies that

$$\delta_{[g_m-g_n,\varphi_n]}(W_n) + \int_{W_n} (g_m-g_n)\varphi_n \, d\sigma(1) = 0.$$

Hence

$$- \delta_{[\varphi_m-\varphi_n,\varphi_n]}(W_n) - \int_{W_n} (\varphi_m-\varphi_n)\varphi_n \, d\sigma(1)$$

$$= \delta_{[g_m-\varphi_m,\varphi_n]}(W_m) - \delta_{[g_n-\varphi_n,\varphi_n]}(W_n)$$

$$+ \int_{W_m} (g_m-\varphi_m)\varphi_n \, d\sigma(1) - \int_{W_n} (g_n-\varphi_n)\varphi_n \, d\sigma(1)$$

$$\leq \delta_{g_m-\varphi_m}(W_m)^{1/2}\cdot\delta_{\varphi_n}(W_m)^{1/2} + \delta_{g_n-\varphi_n}(W_n)^{1/2}\,\delta_{\varphi_n}(W_n)^{1/2}$$

$$+ \{\int_{W_m} (g_m-\varphi_m)^2 \, d|\sigma(1)|\}^{1/2}(\int_{W_m} \varphi_n^2 \, d|\sigma(1)|)^{1/2} +$$

$$+ \{\int_{W_n} (g_n-\varphi_n)^2 \, d|\sigma(1)|\}^{1/2}(\int_{W_n} \varphi_n^2 \, d|\sigma(1)|)^{1/2}$$

$$\leq (\|g_m-\varphi_m\|_{E,W_m} + \|g_n-\varphi_n\|_{E,W_n})\|\varphi_n\|_{E,U}$$

$$\leq (2\beta_U-1)(2^{-m}+2^{-n})I_U(\sigma(\varphi_n))^{1/2} \leq (2\beta_U-1)2^{-n+1}M.$$

Therefore, we have

(6.10) $$0 \leq \|\varphi_n-\varphi_m\|^2_{E,U} \leq (2\beta_U-1)(a_m-a_n+2^{-n+1}M')$$

for $m > n$, where $M' = 2(2\beta_U-1)M$. It follows that $a_{n+1} \geq a_n-2^{-n+1}M'$, i.e.,

$$a_{n+1} + 2^{-n+1}M' \geq a_n + 2^{-n+2}M'.$$

Since $\{a_n+2^{-n+2}M'\}$ is bounded, it follows that it is convergent, and hence $\{a_n\}$ is also convergent. Hence, in view of (6.10), $\|\varphi_n - \varphi_m\|_{E,U} \to 0$ $(n,m \to \infty)$. Since $\|\varphi_n - g_n\|_{E,W_n} \to 0$ $(n \to \infty)$, it follows that $\|g_m - g_n\|_{E,W_n} \to 0$ $(m > n \to \infty)$, so that

$$\|u_m - u_n\|_{E,W_n} \to 0 \qquad (m > n \to \infty).$$

Using Theorem 6.5 , we see that there is $u \in \mathcal{H}(U)$ such that $\|u_m - u\|_{E,W_n} \to 0$ $(m \to \infty)$ for each n. If $\sigma(1) \neq 0$, then Corollary 6.6 implies that $u_n \to u$ locally uniformly. If $\sigma(1) = 0$, then choose $\mu \in \mathcal{M}_I^+(U)$ such that Supp μ is compact, $G_U^\mu \in \mathcal{P}_{BC}(U)$ and $\mu \neq 0$. Let Supp $\mu \subset W_{n_0}$. Then, for $m > n \geq n_0$, using Lemma 4.12 , we have

$$\{\int_{W_n} (g_m-g_n)^2 \, d\mu\}^{1/2}$$

$$\leq \{\int_{W_n} (g_m-\varphi_m)^2 \, d\mu\}^{1/2} + \{\int_{W_n} (\varphi_m-\varphi_n)^2 \, d\mu\}^{1/2} + $$

$$+ \{\int_{W_n} (\varphi_n-g_n)^2 \, d\mu\}^{1/2}$$

$$\leq (\sup G_U^\mu)^{1/2} \{I_{W_m}[\sigma(g_m) - \sigma(\varphi_m)]^{1/2} + I_U[\sigma(\varphi_m) - \sigma(\varphi_n)]^{1/2} +$$

$$+ I_{W_n}[\sigma(\varphi_n) - \sigma(g_n)]^{1/2}\} \to 0 \qquad (m > n \to \infty).$$

Hence, $\int_{W_n} (u_m-u_n)^2 \, d\mu \to 0$ $(m > n \to \infty)$. Then, it follows from Corollary 6.7 that $u_n \to u+c$ (c:const.) locally uniformly on U. Taking u+c instead of u, $u_n \to u$ locally uniformly in this case too. Now, it follows that $g_n \to f - u$ locally uniformly on U, and hence

(6.11) $\quad u + \varphi_n \to f$ locally uniformly on U.

By Lemma 6.3, $\{\|u_m\|_{E,W_m}\}$ is bounded and by Proposition 6.4,

$$\|u\|_{E,W_n} \le \liminf_{m\to\infty} \|u_m\|_{E,W_n} \le \liminf_{m\to\infty} \|u_m\|_{E,W_m}$$

for each n. Hence $\|u\|_{E,U} < +\infty$, i.e., $u \in \mathcal{H}_E(U)$. Thus, $u + \varphi_n \in \mathcal{H}_E(U) + \mathcal{R}_o(U)$. Also, by Proposition 6.4, $\|u_n-u\|_{E,W_n} \le \liminf_{m\to\infty} \|u_n-u_m\|_{E,W_n} \to 0$ $(n \to \infty)$. For any $\varepsilon > 0$, choose n_1 so that

$$\|f\|_{E,U\setminus W_{n_1}} < \frac{\varepsilon}{2} \quad \text{and} \quad \|u\|_{E,U\setminus W_{n_1}} < \frac{\varepsilon}{2}\,.$$

Then, for $m \ge n_1$,

$$\|f - (u+\varphi_m)\|^2_{E,U}$$

$$\le 2\|f\|^2_{E,U\setminus W_m} + 2\|u\|^2_{E,U\setminus W_m} + 2\|u_m - u\|^2_{E,W_m} + 2\,\|g_m - \varphi_m\|^2_{E,W_m}.$$

Hence

$$\limsup_{m\to\infty} \|f - (u+\varphi_m)\|_{E,U} \le \varepsilon.$$

Therefore, $\|f - (u+\varphi_n)\|_{E,U} \to 0$ $(n \to \infty)$ and we have proved (a).

(b) If $f = v + p_1 - p_2$ with $v \in \mathcal{H}(U)$, $p_1, p_2 \in \mathcal{P}(U) \cap \mathcal{C}(U)$, then $v + p_1 \ge f \ge v - p_2$, so that $v + p_1 \ge u_n \ge v - p_2$ for all n. Hence $v + p_1 \ge u \ge v - p_2$, which implies $v = u$. Therefore, $f - u \in \mathcal{Q}_C(U) \cap \mathcal{R}_E(U) = \mathcal{Q}_{EC}(U)$.

(c) If $f \in \mathcal{Q}_C(U)$, then $u = v = 0$ in the above arguments. Hence $\|f - \varphi_n\|_{E,U} \to 0$ $(n \to \infty)$ and $\varphi_n \to f$ locally uniformly on U by (6.11).

Now we obtain extensions of Corollary 4.7 and Theorem 5.1 :

Proposition 6.7. Let U be a PB-domain. Then

$$\delta_g(U) + \int_U g^2\, d\sigma(1) \geq (2\beta_U - 1)^{-1} \|g\|^2_{E,U} \quad \text{for } g \in \mathcal{Q}_{EC}(U)$$

and

$$\delta_{[u,g]}(U) + \int_U ug\, d\sigma(1) = 0 \qquad \text{for } u \in \mathcal{H}_E(U),\ g \in \mathcal{Q}_{EC}(U)$$

Proof. By (c) of the previous proposition, there is a sequence $\{g_n\}$ in $\mathcal{R}_o(U) \subset \mathcal{Q}_{IC}(U)$ such that $\|g_n - g\|_{E,U} \to 0$ $(n \to \infty)$. By Corollary 4.7 and Theorem 5.1,

$$\delta_{g_n}(U) + \int_U g_n^2\, d\sigma(1) = I_U(\sigma(g_n)) \geq (2\beta_U - 1)\|g_n\|^2_{E,U}$$

and

$$\delta_{[u,g_n]}(U) + \int_U ug_n\, d\sigma(1) = 0.$$

Hence, letting $n \to \infty$, we obtain the proposition.

Proposition 6.8. Let U be a PB-domain. If $f = u + g$ with $u \in \mathcal{H}_E(U)$ and $g \in \mathcal{Q}_{EC}(U)$, then

$$\|u\|_{E,U} \leq M(\beta_U)\|f\|_{E,U} \quad \text{and} \quad \|g\|_{E,U} \leq (2\beta_U - 1)\|f\|_{E,U},$$

where $M(t)$ is the same function as in Lemma 6.3.

Proof. Choose, by Proposition 6.6 , (c), $\{g_n\} \subset \mathcal{R}_o(U)$ such that $\|g_n - g\|_{E,U} \to 0$ $(n \to \infty)$ and put $f_n = u + g_n$. Then $\|f_n - f\|_{E,U} \to 0$ $(n \to \infty)$. By Lemma 6.3,

$$\|u\|_{E,U} \leq M(\beta_U)\|f_n\|_{E,U} \text{ and } \|g_n\|_{E,U} \leq (2\beta_U - 1)\|f_n\|_{E,U}.$$

Hence, letting $n \to \infty$, we obtain the required inequalities.

6-6. Density of $\mathcal{H}_{BE}(U)$ in $\mathcal{H}_E(U)$ (cf. [28], [13] for the classical case)

Lemma 6.5. Let U be a PB-domain and μ be a non-negative measure on U having a compact support in U. Then

$$\int_U g^2 \, d\mu \leq (\sup G_U^\mu)\|g\|_{E,U}^2$$

for all $g \in \mathcal{Q}_{EC}(U)$.

Proof. This is a consequence of Lemma 4.12 , Corollary 4.7 and Proposition 6.6 , (c).

Proposition 6.9. Let U be a PB-domain, $u \in \mathcal{H}_E^+(U)$ and $\alpha > 0$. Then

$$\min(u,\alpha) = v_\alpha + g_\alpha$$

with $v_\alpha \in \mathcal{H}_E(U)$ and $g_\alpha \in \mathcal{Q}_{EC}(U)$.

Proof. By Corollary 3.2, $\delta_{\min(u,\alpha)} \leq \delta_u$. Since $0 \leq \min(u,\alpha) \leq u$, it follows that $\|\min(u,\alpha)\|_{E,U} \leq \|u\|_{E,U} < +\infty$. Hence, $\min(u,\alpha) \in \mathcal{R}_E(U)$. Since we can write

$$\min(u,\alpha) = \min(u + \alpha G_U^{\sigma(1)^-}, \alpha s_U) - \alpha G_U^{\sigma(1)^-}$$

and $\min(u + \alpha G_U^{\sigma(1)^-}, \alpha s_U) \in \mathcal{S}_C(U)$, $\min(u,\alpha) \in \mathcal{H}(U) + \mathcal{Q}_C(U)$. Hence, by Proposition 6.6 , (b), $\min(u,\alpha) \in \mathcal{H}_E(U) \oplus \mathcal{Q}_{EC}(U)$.

Proposition 6.10. Let U be a PB-domain. Given $u \in \mathcal{H}_E^+(U)$, let v_α be as in the previous proposition for each $\alpha > 0$. Then $v_\alpha \to u$ locally uniformly on U and $\|v_\alpha - u\|_{E,U} \to 0$ $(\alpha \to \infty)$.

Proof. Clearly, $v_\alpha \le v_{\alpha'} \le u$ for $0 < \alpha < \alpha'$. Hence

$$v = \lim_{\alpha\to\infty} v_\alpha$$

exists, $v \in \mathcal{H}(U)$ and $v \le u$. Furthermore, $v_\alpha \to v$ locally uniformly on U. Let

$$A_\alpha = \{x \in U \mid u(x) \ge \alpha\} \qquad \text{for } \alpha > 0.$$

Since $\max(u,\alpha) = \alpha$ on the open set $U \setminus A_\alpha$, $\delta_{\max(u,\alpha)} = 0$ on $U \setminus A_\alpha$. Hence

$$\delta_{u-\min(u,\alpha)} = \delta_{\max(u,\alpha)} = \chi_{A_\alpha}\, \delta_{\max(u,\alpha)},$$

so that

$$\delta_{u-\min(u,\alpha)}(U) = \delta_{\max(u,\alpha)}(A_\alpha) \le \delta_u(A_\alpha) \to 0 \ (\alpha \to \infty).$$

On the other hand, Lebesgue's convergence theorem implies

$$\int_U \{u - \min(u,\alpha)\}^2 \, d|\sigma(1)| \to 0 \qquad (\alpha \to \infty).$$

Hence

(6.12) $$\|u - \min(u,\alpha)\|_{E,U} \to 0 \qquad (\alpha \to \infty).$$

By the previous proposition

$$\min(u,\alpha) - \min(u,\alpha') = v_\alpha - v_{\alpha'} + g_{\alpha,\alpha'}$$

with $g_{\alpha,\alpha'} \in \mathcal{Q}_{EC}(U)$. By Proposition 6.8,

$$\|v_\alpha - v_{\alpha'}\|_{E,U} \le M(\beta_U)\|\min(u,\alpha) - \min(u,\alpha')\|_{E,U}$$

$$\to 0 \ (\alpha, \alpha' \to \infty).$$

Since $v_\alpha \to v$, it follows from Theorem 6.5 and Corollary 6.6 that $v \in \mathcal{H}_E(U)$ and

(6.13) $$\|v - v_\alpha\|_{E,U} \to 0 \qquad (\alpha \to \infty).$$

Let $\min(u,\alpha) = v_\alpha + g_\alpha$ with $g_\alpha \in \mathcal{Q}_{EC}(U)$. By (6.12) and (6.13), we have

(6.14) $$\|g_\alpha - u + v\|_{E,U} \to 0 \qquad (\alpha \to \infty).$$

By Proposition 6.7,

$$\delta_{[u-v,g_\alpha]}(U) + \int_U (u-v)g_\alpha \, d\sigma(1) = 0.$$

Hence, in view of (6.14), we have

$$\delta_{g_\alpha}(U) + \int_U g_\alpha^2 \, d\sigma(1) \to 0 \quad (\alpha \to \infty).$$

Then, by Proposition 6.7 , we see that $\|g_\alpha\|_{E,U} \to 0$ $(\alpha \to \infty)$. Now, let μ be a non-zero non-negative measure on U having a compact support in U. Then, by Lemma 6.5, $\int_U g_\alpha^2 \, d\mu \to 0$ $(\alpha \to \infty)$. Hence, for some $\alpha_n \to \infty$ and some $x \in U$, $g_{\alpha_n}(x) \to 0$ $(n \to \infty)$, so that $u(x) = v(x)$ for some $x \in U$. Since $v \leq u$ and $v, u \in \mathcal{H}(U)$, it follows that $u = v$. Hence $v_\alpha \to u$ locally uniformly and $\|v_\alpha - u\|_{E,U} \to 0$ $(\alpha \to \infty)$.

Corollary 6.9. If U is a PB-domain, then $\mathcal{H}_{BE}(U)$ is dense in $\mathcal{H}_E(U)$.

Proof. First let $u \in \mathcal{H}_E^+(U)$ and let $\min(u,\alpha) = v_\alpha + g_\alpha$ with $v_\alpha \in \mathcal{H}_E(U)$ and $g_\alpha \in \mathcal{Q}_{EC}(U)$. Since $\min(u,\alpha) \leq \alpha s_U$, $v_\alpha \leq \alpha s_U$, so that v_α is bounded, i.e., $v_\alpha \in \mathcal{H}_{BE}(U)$. Hence, by Proposition 6.9, u is approximated by functions in $\mathcal{H}_{BE}(U)$. If $u \in \mathcal{H}_E(U)$ is arbitrary, then $u = u_1 - u_2$ with $u_1, u_2 \in \mathcal{H}_E^+(U)$ by virtue of

Theorem 6.2. Since u_1, u_2 are approximated by functions in $\mathcal{H}_{BE}(U)$, so is u.

Corollary 6.10. Let U be a PB-domain. If $\mathcal{H}_E(U)$ contains non-constant functions, then it contains a non-constant bounded function.

Results similar to Proposition 6.10 and its corollaries can be obtained for $\mathcal{H}_D(U)$ in case $\sigma(1) \geq 0$ on U. We first prove

Lemma 6.6. Let U be a P-domain and suppose $\sigma(1) \geq 0$ on U. Let $u \in \mathcal{H}_D^+(U)$ and $\alpha > 0$. Then the greatest harmonic minorant $v_\alpha = u \underset{U}{\wedge} \alpha$ of u and α belongs to $\mathcal{H}_D(U)$ and

$$\|u - v_\alpha\|_{D,U} \leq \|\max(u,\alpha)\|_{D,U}.$$

Proof. Let $\{W_n\}$ be an exhaustion of U and put $w_n = u \underset{W_n}{\wedge} \alpha$.

Then, $v_\alpha \leq w_n \leq \min(u,\alpha) \leq u$ on W_n and $w_n \downarrow v_\alpha$ $(n \to \infty)$. By Lemma 5.7, $w_n \in \mathcal{H}_E(W_n)$ and $\min(u,\alpha)|W_n - w_n \in \mathcal{P}_{IC}(W_n)$. Also, $u|W_n \in \mathcal{H}_E(W_n)$. Hence, by Theorem 5.1,

$$\delta_{[w_n,\min(u,\alpha)-w_n]}(W_n) + \int_{W_n} w_n\{\min(u,\alpha) - w_n\}\, d\sigma(1) = 0$$

and

$$\delta_{[u,\min(u,\alpha)-w_n]}(W_n) + \int_{W_n} u\{\min(u,\alpha) - w_n\}\, d\sigma(1) = 0.$$

Since

$$\int_{W_n} w_n\{\min(u,\alpha) - w_n\} d\sigma(1) \leq \int_{W_n} u\{\min(u,\alpha) - w_n\}\, d\sigma(1),$$

we have

$$\delta_{[w_n,\min(u,\alpha)-w_n]}(W_n) \geq \delta_{[u,\min(u,\alpha)-w_n]}(W_n).$$

Hence

$$\delta_{w_n-u}(W_n) \leq - \delta_{[w_n-u,u-\min(u,\alpha)]}(W_n)$$

$$\leq \delta_{w_n-u}(W_n)^{1/2}\ \delta_{\max(u,\alpha)}(W_n)^{1/2},$$

so that

$$\|w_n-u\|^2_{D,W_n} = \delta_{w_n-u}(W_n) \leq \delta_{\max(u,\alpha)}(W_n) \leq \delta_{\max(u,\alpha)}(U) =$$

$$= \|\max(u,\alpha)\|^2_{D,U}.$$

Fix m. Since $w_n \to v_\alpha$ uniformly on W_m, Proposition 6.4 implies that

$$\|v_\alpha-u\|_{D,W_m} \leq \liminf_{n\to\infty} \|w_n-u\|_{D,W_m}$$

$$\leq \liminf_{n\to\infty} \|w_n-u\|_{D,W_n} \leq \|\max(u,\alpha)\|_{D,U}.$$

Letting $m\to\infty$, we obtain the required inequality.

<u>Proposition 6.11.</u> Let U be a P-domain and suppose $\sigma(1) \geq 0$ on U. Let $u \in \mathcal{H}^+_D(U)$ and $v_\alpha = u \underset{U}{\wedge} \alpha$ for $\alpha>0$. Then $v_\alpha \to u$ locally uniformly on U and $\|u - v_\alpha\|_{D,U} \to 0$ as $\alpha \to \infty$.

<u>Proof.</u> Let $A_\alpha = \{x\in U \mid u(x) \geq \alpha\}$. Then

$$\|\max(u,\alpha)\|^2_{D,U} = \delta_{\max(u,\alpha)}(A_\alpha) \leq \delta_u(A_\alpha) \to 0 \qquad (\alpha \to \infty).$$

Hence by the above lemma $\|u - v_\alpha\|_{D,U} \to 0$ $(\alpha \to \infty)$. If $\sigma(1) \not\equiv 0$, then it follows from Corollary 6.6 that $v_\alpha \to u$ locally uniformly. In case $\sigma(1) = 0$, Corollary 6.7 implies that $v_\alpha \to u-c$ locally uniformly on U with a constant $c \geq 0$, since $v_\alpha \leq v_{\alpha'} \leq u$ for $0 < \alpha < \alpha'$. Let $\tilde{v}_\alpha$ be the greatest harmonic minorant of $\min(u-c,\alpha)$. Since $v_\alpha \leq \min(u-c,\alpha)$, we have

$v_\alpha \leq \tilde{v}_\alpha$. On the other hand, since $u-c \leq u$, $\tilde{v}_\alpha \leq v_\alpha$. Thus $\tilde{v}_\alpha = v_\alpha$. Since

$$\min(u-c,\alpha) = \min(u,\alpha+c) - c,$$

$v_\alpha = \tilde{v}_\alpha = v_{\alpha+c} - c$. Letting $\alpha \to \infty$, we obtain $u-c = u$, i.e., $c = 0$. Thus $v_\alpha \to u$ locally uniformly on U in this case too.

Corollary 6.11. If U is a P-domain such that $\sigma(1) \geq 0$ on U, then $\mathcal{H}_{BD}(U)$ is dense in $\mathcal{H}_D(U)$.

Proof. This is seen in the same way as Corollary 6.9 using Corollary 6.2.

Corollary 6.12. Let U be a P-domain such that $\sigma(1) \geq 0$ on U. If $\mathcal{H}_D(U)$ contains non-constant functions, then it contains a non-constant bounded function.

Open question: Do similar results hold for $\mathcal{H}_{D'}(U)$ or $\mathcal{H}_D(U)$ in case $\sigma(1)^- \not\equiv 0$?

§7. Functional completion

7-1. Completion of $\mathcal{R}(U)$ (cf. [25])

Lemma 7.1. Let $U \in \mathcal{O}_X$ and $f_n \in \mathcal{R}(U)$, $n = 1,2,\ldots$. If $f_n \to 0$ locally uniformly on U and if

$$\delta_{f_n - f_m}(K) \to 0 \qquad (n,m \to \infty)$$

for any compact set K in U, then $\delta_{f_n}(K) \to 0$ $(n \to \infty)$ for any compact set K in U.

Proof. Let V be a PC-domain such that $\overline{V} \subset U$ and there is $h \in \mathcal{H}(V')$ with $V' \supset V$ and

$$0 < \inf_V h \leq \sup_V h < +\infty.$$

Consider the harmonic space $(V', \mathcal{H}_{V',h})$. Then $\sigma^{(h)}(1) = 0$ on V'. Since $\delta_f^{(h)} = h^2\delta_f$ for the gradient measures $\delta_f^{(h)}$ associated with $\sigma^{(h)}$, $\delta^{(h)}_{f_n-f_m}(V) \to 0$ $(n,m \to \infty)$ and if we can show that $\delta^{(h)}_{f_n}(V) \to 0$, then we have $\delta_{f_n}(V) \to 0$. Hence we may assume that $\sigma(1) = 0$ on V. Let $f_n = u_n + p_n$ with $u_n \in \mathcal{H}_E(V)$ and $p_n \in \mathcal{Q}_{IC}(V)$ (Lemma 5.7). By Theorem 5.1, $\delta_{[u_n,p_m]}(V) = 0$ for any n,m, so that

$$\delta_{f_n}(V) = \delta_{u_n}(V) + \delta_{p_n}(V)$$

and

$$\delta_{f_n-f_m}(V) = \delta_{u_n-u_m}(V) + \delta_{p_n-p_m}(V).$$

It follows that $\delta_{u_n-u_m}(V) \to 0$ and $\delta_{p_n-p_m}(V) \to 0$ $(n,m \to \infty)$. Since $f_n \to 0$ uniformly on V, we see that $u_n \to 0$ uniformly on V, and hence $p_n \to 0$ uniformly on V. By Proposition 6.4,

$$\delta_{u_n}(V) \le \liminf_{m\to\infty} \delta_{u_n-u_m}(V) \to 0 \qquad (n \to \infty),$$

i.e.,

(7.1) $$\delta_{u_n}(V) \to 0 \qquad (n \to \infty).$$

On the other hand, since $|\sigma(p_n)|(V) = |\sigma(f_n)|(V) < +\infty$, we have by Theorem 4.3

$$\delta_{p_n}(V) = \int_V p_n \, d\sigma(p_n) = \lim_{m\to\infty} \int_V (p_n - p_m)\, d\sigma(p_n)$$

$$= \lim_{m\to\infty} \delta_{[p_n-p_m,p_n]}(V) \le \delta_{p_n}(V)^{1/2} \liminf_{m\to\infty} \delta_{p_n-p_m}(V)^{1/2}.$$

Hence

$$\delta_{p_n}(V) \le \liminf_{m\to\infty} \delta_{p_n - p_m}(V) \to 0 \qquad (n \to \infty).$$

Thus, together with (7.1), we see that $\delta_{f_n}(V) \to 0 \quad (n \to \infty)$. Now, any compact set K in U can be covered by a finite number of PC-sets V as above. Hence we obtain the lemma.

For $U \in \mathcal{O}_X$, let $\mathcal{D}(U)$ be the set of all $f \in \mathbf{C}(U)$ for which there is a sequence $\{f_n\}$ in $\mathcal{R}(U)$ such that $f_n \to f$ locally uniformly on U and $\delta_{f_n - f_m}(K) \to 0$ $(n,m \to \infty)$ for any compact set K in U. Given $f \in \mathcal{D}(U)$, any sequence $\{f_n\}$ in $\mathcal{R}(U)$ as above will be called a <u>defining sequence</u> of f. If $\{\tilde{f}_n\}$ is another defining sequence of f, then Lemma 7.1 implies that $\delta_{f_n - \tilde{f}_n}(K) \to 0$ for any compact set K in U, so that $\lim_{n\to\infty} \delta_{f_n}$ and $\lim_{n\to\infty} \delta_{\tilde{f}_n}$ define the same non-negative measure on U. Hence we define

(7.2) $$\delta_f = \lim_{n\to\infty} \delta_{f_n}$$

for $f \in \mathcal{D}(U)$ with a defining sequence $\{f_n\}$ of f. If $f,g \in \mathcal{D}(U)$, then we define

(7.3) $$\delta_{[f,g]} = \lim_{n\to\infty} \delta_{[f_n,g_n]}$$

with defining sequences $\{f_n\}$ and $\{g_n\}$ of f and g, respectively. It is a well-defined signed measure on U. Obviously, $\mathcal{R}(U) \subset \mathcal{D}(U)$ and δ_f, $\delta_{[f,g]}$ thus defined coincide with old ones for $f,g \in \mathcal{R}(U)$. It is easy to see that $\mathcal{D}(U)$ is a linear space, $(f,g) \to \delta_{[f,g]}$ is bilinear on $\mathcal{D}(U) \times \mathcal{D}(U)$ and $\delta_f = \delta_{[f,f]}$, so that

$$\delta_{f+g} = \delta_f + 2\delta_{[f,g]} + \delta_g$$

for $f,g \in \mathcal{D}(U)$. It is also easy to see that $\mathcal{D}: U \to \mathcal{D}(U)$ is a sheaf and the definition of $\delta_{[f,g]}$ does not depend on U.

<u>Proposition 7.1.</u> Let $U \in \mathcal{O}_X$

(a) If $\{f_n\}$ is a defining sequence of $f \in \mathcal{D}(U)$, then

$$\delta_{f_n - f}(K) \to 0 \qquad (n \to \infty)$$

for any compact set K in U.

(b) $\mathcal{D}(U)$ is an algebra and

$$\delta_{[fg,\varphi]} = f\delta_{[g,\varphi]} + g\delta_{[f,\varphi]}$$

for any $f,g,\varphi \in \mathcal{D}(U)$.

<u>Proof.</u> (a) Since $\{f_n - f_m\}_m$ is a defining sequence of $f_n - f$ for each n,

$$\delta_{f_n - f}(K) = \lim_{m\to\infty} \delta_{f_n - f_m}(K) \to 0 \qquad (n \to \infty)$$

for any compact set K in U.

(b) Let $f,g \in \mathcal{D}(U)$ and $\{f_n\}$, $\{g_n\}$ be the respective defining sequences. Then $f_n g_n \in \mathcal{R}(U)$ and $f_n g_n \to fg$ locally uniformly on U. For any compact set K in U,

$$\begin{aligned}
&\delta_{f_n g_n - f_m g_m}(K) \\
&\leq 2\{\delta_{(f_n - f_m)g_n}(K) + \delta_{f_m(g_n - g_m)}(K)\} \\
&= 2\{\int_K g_n^2 \, d\delta_{f_n - f_m} + \int_K (f_n - f_m)^2 \, d\delta_{g_n} \\
&\quad + 2\int_K (f_n - f_m) g_n \, d\delta_{[f_n - f_m, g_n]} \\
&\quad + \int_K f_m^2 \, d\delta_{g_n - g_m} + \int_K (g_n - g_m)^2 \, d\delta_{f_m} \\
&\quad + 2\int_K f_m(g_n - g_m) \, d\delta_{[f_m, g_n - g_m]}\}.
\end{aligned}$$

Since $\{f_n\},\{g_n\}$ are uniformly bounded and uniformly convergent on K, $\{\delta_{f_n}(K)\}$ and $\{\delta_{g_n}(K)\}$ are bounded and $\delta_{f_n-f_m}(K) \to 0$, $\delta_{g_n-g_m}(K) \to 0$ $(n,m \to \infty)$, we see that the last expression above tends to 0 as $n,m \to \infty$. Hence $\{f_n g_n\}$ is a defining sequence of fg, so that $fg \in \mathcal{D}(U)$. Therefore, $\mathcal{D}(U)$ is an algebra. Now, the equality in (b) is easily seen by considering also the defining sequence of φ.

For $f \in \mathcal{D}(U)$ and a compact set K in U let

$$P_K(f) = \delta_f(K)^{1/2} + \sup_K |f|.$$

Then P_K is a semi-norm on $\mathcal{D}(U)$. Proposition 7.1, (a), shows that $\mathcal{D}(U)$ is a completion of $\mathcal{R}(U)$ with respect to the system of semi-norms $\{P_K\}_{K:\text{compact} \subset U}$. Thus, $\mathcal{D}(U)$ is complete with respect to this system of semi-norms.

<u>Proposition 7.2.</u> Let U be a domain in X. If $f \in \mathcal{D}(U)$ and $\delta_f = 0$ on U, then f = const.

<u>Proof.</u> As in the proof of Lemma 7.1, it is enough to consider the case $\sigma(1) = 0$ on U. Let $\{f_n\}$ be a defining sequence of f. Let V be a PC-domain such that $\overline{V} \subset U$ and let $f_n = u_n + p_n$ with $u_n \in \mathcal{H}_E(V)$ and $p_n \in \mathcal{Q}_{IC}(V)$ (Lemma 5.7).
Since $\delta_{f_n}(V) = \delta_{u_n}(V) + \delta_{p_n}(V)$ and $\delta_{f_n}(V) \to \delta_f(V) = 0$, we see that $\delta_{u_n}(V) \to 0$ and $\delta_{p_n}(V) \to 0$ $(n \to \infty)$. Also, since $\{f_n\}$ is uniformly convergent on V, $\{u_n\}$, and hence $\{p_n\}$, is uniformly convergent on V. Let

$$u = \lim_{n\to\infty} u_n \quad \text{and} \quad p = \lim_{n\to\infty} p_n.$$

Then $f = u + p$ on V. By Proposition 6.4, we see that $\delta_u = 0$ on V. Hence, by Theorem 5.4, u = const. on V. On the other hand, since $\delta_{p_n}(V) \to 0$, $\delta_{[p_n,g]}(V) \to 0$ for any $g \in \mathcal{Q}_{IC}(V)$. Hence, by Theorem 4.3,

$$\int_V p_n \, d\sigma(g) \to 0 \qquad (n \to \infty),$$

so that $\int_V p \, d\,\sigma(g) = 0$ for any $g \in \mathcal{Q}_{IC}(V)$ such that $|\sigma(g)|(V) < +\infty$.

Since for any non-empty open set $W \subset V$ we can find $g \in \mathcal{P}_{IC}(U)$ such that $g \not\equiv 0$ and Supp $\sigma(g) \subset W$, it follows that $p = 0$ on V. Hence $f = u =$ const. on V. Since U is covered by such V's and U is connected, $f =$ const. on U.

7-2. Extension of Green's formula and its applications

Lemma 7.2. Let $U \in \mathcal{O}_X$ and $g \in \mathcal{D}_0(U) = \mathcal{D}(U) \cap \mathcal{C}_0(U)$. Then, for any open set U' such that Supp $g \subset U' \subset U$, we can find a defining sequence $\{g_n\}$ of g such that Supp g_n is compact and contained in U' for all n; in particular, $g_n \to g$ uniformly on U.

Proof. By Proposition 2.17 , there is $\varphi \in \mathcal{R}(U)$ such that $0 \le \varphi \le 1$ on U, $\varphi = 1$ on Supp g and Supp φ is compact and contained in U'. Let $\{\tilde{g}_n\}$ be an arbitrary defining sequence of g. Put $g_n = \varphi\tilde{g}_n$. Then $g_n \in \mathcal{R}(U)$, Supp $g_n \subset$ Supp $\varphi \subset U'$ and $g_n \to g$ uniformly on U. Furthermore,

$$\delta_{g_n-g_m}(K) = \delta_{\varphi(\tilde{g}_n-\tilde{g}_m)}(K)$$
$$\le 2\{\int_K \varphi^2 \, d\delta_{\tilde{g}_n-\tilde{g}_m} + \int_K (\tilde{g}_n-\tilde{g}_m)^2 \, d\delta_\varphi\} \to 0$$

$(n,m \to \infty)$ for any compact set K in U. Hence $\{g_n\}$ is a required defining sequence of g.

Theorem 7.1. (Green's formula) Let $U \in \mathcal{O}_X$. If $f \in \mathcal{R}(U)$ and $g \in \mathcal{D}_0(U) (= \mathcal{D}(U) \cap \mathcal{C}_0(U))$, then

$$\delta_{[f,g]}(U) + \int_U fg\, d\sigma(1) = \int_U g\, d\sigma(f).$$

Proof. Let U' be a relatively compact open set such that $\mathrm{Supp}\, g \subset U'$ and $\overline{U}' \subset U$. By the above lemma, we can choose a defining sequence $\{g_n\}$ of g such that $\mathrm{Supp}\, g_n \subset U'$ for all n.

By Theorem 5.3,

$$\delta_{[f,g_n]}(U) + \int_U fg_n\, d\sigma(1) = \int_U g_n\, d\sigma(f)$$

for each n. Letting $n \to \infty$, we easily obtain the required formula.

Proposition 7.3. Let $U \in \mathcal{O}_X$, $f_1, \ldots, f_k \in \mathcal{D}(U)$ and put $\vec{f} = (f_1, \ldots, f_k)$. Let Ω be an open subset of $\mathbb{R}^k$ containing $\vec{f}(U)$. Then, for any $\varphi \in C^1(\Omega)$, $\varphi\circ\vec{f} \in \mathcal{D}(U)$ and

$$\delta_{[\varphi\circ\vec{f},g]} = \sum_{j=1}^{k} \left(\frac{\partial\varphi}{\partial x_j} \circ\vec{f}\right) \delta_{[f_j,g]} \tag{7.4}$$

for any $g \in \mathcal{D}(U)$.

Proof. If $\varphi = \text{const.}$, then $\varphi\circ\vec{f} = \text{const.}$, so that both sides of (7.4) vanish. If $\varphi(x) = x_j$, then both sides of (7.4) are reduced to $\delta_{[f_j,g]}$ $(j=1,\ldots,k)$. Thus, in view of Proposition 7.1 , (b), we see that (7.4) holds for any polynomial φ on $\mathbb{R}^k$ (cf. the proof of Theorem 3.3). If $\varphi \in C^1(\Omega)$, then there is a sequence $\{\varphi_n\}$ of polynomials on $\mathbb{R}^k$ such that $\varphi_n \to \varphi$ and $\partial\varphi_n/\partial x_j \to \partial\varphi/\partial x_j$, $j = 1,\ldots,k$, all locally uniformly on Ω. Since $\vec{f}$ is continuous, it follows that $\varphi_n\circ\vec{f} \to \varphi\circ\vec{f}$ and $(\partial\varphi_n/\partial x_j)\circ\vec{f} \to (\partial\varphi/\partial x_j)\circ\vec{f}$, $j = 1,\ldots,k$, all locally uniformly on U. Then from (7.4) for φ_n, we see that $\{\varphi_n\circ\vec{f}\}$ is a Cauchy sequence with respect to the system $\{P_K\}$. Since $\mathcal{D}(U)$ is complete, it follows that $\varphi\circ\vec{f} \in \mathcal{D}(U)$ and (7.4) holds for the given φ.

As an application of Theorem 7.1 and Proposition 7.3, we obtain the following in case X is a subdomain of $\mathbb{R}^k$:

Theorem 7.2. (cf. Theorem 3.4) Suppose that the base space X of the given harmonic space is a domain in $\mathbb{R}^k$ ($k \geq 1$). If the coordinate functions $x_1, \dots, x_k$ belong to $\mathcal{D}(X)$, then $\mathcal{C}^1(U) \subset \mathcal{D}(U)$ for any $U \in \mathcal{O}_X$ and

(7.5) $$\delta_{[f,g]} = \sum_{i,j=1}^{k} \frac{\partial f}{\partial x_i} \frac{\partial g}{\partial x_j} \alpha_{ij}$$

for $f, g \in \mathcal{C}^1(U)$, where $\alpha_{ij} = \delta_{[x_i, x_j]}$. The matrix (α_{ij}) is symmetric, and positive definite in the following sense: for each $\xi = (\xi_1, \dots, \xi_k) \in \mathbb{R}^k$ with $\xi \neq 0$, $\mu_\xi = \Sigma_{i,j}\, \xi_i \xi_j \alpha_{ij}$ is a positive measure on X whose support is the whole space X.
Furthermore, if $f \in \mathcal{C}^1(U) \cap \mathcal{R}(U)$, then

(7.6) $$\sum_{i,j=1}^{k} \int_U \frac{\partial f}{\partial x_i} \frac{\partial \varphi}{\partial x_j}\, d\alpha_{ij} + \int_U f\varphi\, d\sigma(1) = \int_U \varphi\, d\sigma(f)$$

for all $\varphi \in \mathcal{C}_0^1(U)$; thus, for $u \in \mathcal{C}^1(U) \cap \mathcal{R}(U)$, $u \in \mathcal{H}(U)$ if and only if

(7.7) $$\sum_{i,j=1}^{k} \int_U \frac{\partial u}{\partial x_i} \frac{\partial \varphi}{\partial x_j}\, d\alpha_{ij} + \int_U u\varphi\, d\sigma(1) = 0$$

for all $\varphi \in \mathcal{C}_0^1(U)$.

Proof. The assertion $\mathcal{C}^1(U) \subset \mathcal{D}(U)$ and the equation (7.5) are immediate consequences of Proposition 7.3. Since

$$\mu_\xi = \sum_{i,j} \xi_i \xi_j \delta_{[x_i, x_j]} = \delta_{\sum_i \xi_i x_i},$$

$\mu_\xi \geq 0$ and $\mu_\xi = 0$ on a domain $U \subset X$ implies $\Sigma_i \xi_i x_i = \text{const.}$, which is impossible unless $U = \emptyset$. Hence (α_{ij}) is positive definite. Since $\mathcal{C}_0^1(U) \subset \mathcal{D}_0(U)$, (7.6) follows from Theorem 7.1

and (7.5). Thus, if $u \in \mathcal{C}^1(U) \cap \mathcal{H}(U)$, then (7.7) holds. Conversely, if $u \in \mathcal{C}^1(U) \cap \mathcal{R}(U)$ satisfies (7.7), then $\int_U \varphi \, d\sigma(u) = 0$ for all $\varphi \in \mathcal{C}_0^1(U)$ by (7.6). Hence $\sigma(u) = 0$ on U, so that $u \in \mathcal{H}(U)$.

As an application of Proposition 7.3 , we have

<u>Proposition 7.4.</u> $\mathcal{D}(U)$ is a vector lattice with respect to the natural order; in fact, $f \in \mathcal{D}(U)$ implies $|f| \in \mathcal{D}(U)$ and $\delta_{|f|} = \delta_f$. Furthermore, $\delta_f(f^{-1}(0)) = 0$ for any $f \in \mathcal{D}(U)$.

<u>Proof.</u> Let $\{\varphi_n\}$ be a sequence of $\mathcal{C}^1$-functions on $\mathbb{R}$ such that $\varphi_n(-t) = -\varphi_n(t)$, $0 \le \varphi_n'(t) \le 1$ for all $t \in R$, $\varphi_n = 0$ on a neighborhood of 0 and $\varphi_n'(t) \uparrow 1$ $(n \to \infty)$ if $t > 0$. Then $\varphi_n(t) \to t$ and $|\varphi_n(t)| \to |t|$ locally uniformly on $\mathbb{R}$. Hence $\varphi_n \circ f \to f$ and $|\varphi_n| \circ f \to |f|$ both locally uniformly on U. Furthermore, $|\varphi_n| \in \mathcal{C}^1(\mathbb{R})$ for each n. By Proposition 7.3, $\varphi_n \circ f \in \mathcal{D}(U)$, $|\varphi_n| \circ f \in \mathcal{D}(U)$ for each n and

$$\delta_{\varphi_n \circ f - \varphi_m \circ f} = \delta_{|\varphi_n| \circ f - |\varphi_m| \circ f} = [(\varphi_n' - \varphi_m') \circ f]^2 \delta_f .$$

Since $|\varphi_n'| \le 1$, Lebesgue's convergence theorem implies that

$$\delta_{\varphi_n \circ f - \varphi_m \circ f}(K) = \delta_{|\varphi_n| \circ f - |\varphi_m| \circ f}(K) \to 0 \qquad (n,m \to \infty)$$

for any compact set K in U. Since $\mathcal{D}(U)$ is complete with respect to $\{P_K\}$, it follows that $|f| \in \mathcal{D}(U)$ and

$$\delta_{|f|}(A) = \lim_{n\to\infty} \delta_{|\varphi_n| \circ f}(A) = \lim_{n\to\infty} \delta_{\varphi_n \circ f}(A) = \delta_f(A)$$

for any relatively compact Borel set A such that $\overline{A} \subset U$, which implies $\delta_{|f|} = \delta_f$. If $A = f^{-1}(0) \cap K$ with a compact set K in U, then $\varphi_n' \circ f = 0$ on A, and hence

$$\delta_{\varphi_n \circ f}(f^{-1}(0) \cap K) = \int_{f^{-1}(0) \cap K} (\varphi_n' \circ f)^2 \, d\delta_f = 0.$$

Therefore, $\delta_f(f^{-1}(0)\cap K) = 0$ for any compact set K in U, so that $\delta_f(f^{-1}(0)) = 0$.

Corollary 7.1. (a) For $f \in \mathcal{D}(U)$

$$\delta_{\max(f,0)} = \chi_{\{f>0\}}\delta_f \text{ and } \delta_{\min(f,0)} = \chi_{\{f<0\}}\delta_f.$$

(b) For $f,g \in \mathcal{D}(U)$

$$\delta_f|\{f=g\} = \delta_g|\{f=g\} = \delta_{[f,g]}|\{f=g\}.$$

Proof. (a) is an immediate consequence of Proposition 7.4.

(b) Let A be a relatively compact Borel set contained in $\{f = g\}$. By Proposition 7.4, $\delta_{f-g}(A) = 0$. Since

$$|\delta_f(A)^{1/2} - \delta_g(A)^{1/2}| \leq \delta_{f-g}(A)^{1/2},$$

we have $\delta_f(A) = \delta_g(A)$. Hence $\delta_f|\{f=g\} = \delta_g|\{f=g\}$. Since

$$\delta_{f-g} = \delta_f + \delta_g - 2\delta_{[f,g]},$$

it also follows that $\delta_{[f,g]}|\{f=g\} = \delta_f|\{f=g\}$.

7-3. Spaces $\mathcal{D}_E(U)$, $\mathcal{D}_D(U)$, $\mathcal{D}_{E,o}(U)$ and $\mathcal{D}_{D,o}(U)$

For an open set U in X, let

$$\mathcal{D}_E(U) = \{f \in \mathcal{D}(U) \mid \delta_f(U) + \int_U f^2 \, d|\sigma(1)| < +\infty\}$$

and

$$\mathcal{D}_D(U) = \{f \in \mathcal{D}(U) \mid \delta_f(U) < +\infty\}.$$

Obviously, these are linear spaces and $\mathcal{D}_E(U) \subset \mathcal{D}_D(U)$; $\mathcal{D}_E(U) = \mathcal{D}_D(U)$ if $\sigma(1) = 0$ on U.

From now on, we consider only the case U is a domain. If U is an open set, we may discuss on each component of U. If $\sigma(1) \neq 0$ on U, then we consider the norm $\|\cdot\|_{E,U}$ defined by

$$\|f\|_{E,U}^2 = \delta_f(U) + \int_U f^2 \, d|\sigma(1)|$$

on $\mathcal{D}_E(U)$. For a norm on $\mathcal{D}_D(U)$, we choose $\mu \in \mathfrak{M}_{C,o}^+(U)$, $\mu \neq 0$, and

consider

$$\|f\|_{D,\mu,U}^2 = \delta_f(U) + \int_U f^2 \, d\mu \qquad , \ f \in \mathcal{D}_D(U),$$

where

$$\mathfrak{M}_{C,o}^+(U) = \{\mu \in \mathfrak{M}_C^+(U) \mid \text{Supp } \mu \text{ is compact in } U\}$$

and

$$\mathfrak{M}_C^+(U) = \{\mu \in \mathfrak{M}^+(U) \mid G_V^\mu \text{ is continuous for any PC-domain } V \text{ with } \overline{V} \subset U\}.$$

Since $\delta_{|f|} = \delta_f$ for $f \in \mathcal{D}(U)$, $\mathcal{D}_E(U)$ and $\mathcal{D}_D(U)$ are closed under max. and min. operations and

$$\| |f| \|_{E,U} = \|f\|_{E,U} \qquad , \qquad \| |f| \|_{D,\mu,U} = \|f\|_{D,\mu,U}.$$

<u>Lemma 7.3.</u> Let U be a domain in X, $\mu \in \mathfrak{M}_{C,o}^+(U)$ and $\nu \in \mathfrak{M}_C^+(U)$. If $\nu \neq 0$, then there is $M = M(\mu,\nu) > 0$ such that

$$\int_U f^2 \, d\mu \leq M(\delta_f(U) + \int_U f^2 \, d\nu)$$

for all $f \in \mathcal{D}(U)$.

<u>Proof.</u> It is enough to show that if $\{f_n\}$ is a sequence in $\mathcal{D}(U)$ such that

$$\delta_{f_n}(U) + \int_U f_n^2 \, d\nu \to 0 \qquad (n \to \infty),$$

then $\int_U f_n^2\, d\mu \to 0$ $(n \to \infty)$. Let U' be a relatively compact domain such that $\mathrm{Supp}\,\mu \subset U'$, $\overline{U'} \subset U$ and $\nu(U') > 0$. By the definition of $\mathcal{D}(U)$, we can find $g_n \in \mathcal{R}(U)$ such that

$$\delta_{f_n - g_n}(U') + \int_{U'} (f_n - g_n)^2\, d\nu < \frac{1}{n}$$

and

$$\int_{U'} (f_n - g_n)^2\, d\mu < \frac{1}{n}$$

for each n. Then

$$\delta_{g_n}(U') + \int_{U'} g_n^2\, d\nu \to 0 \qquad (n \to \infty).$$

Let V be any domain in U' for which there exists $h \in \mathcal{H}(V)$ with $h > 0$ on V, and consider the harmonic space $(V, \mathcal{H}_{V,h})$. For any domain W such that $\overline{W} \subset V$, $g_n = u_n + p_n$ on W with $u_n \in \mathcal{H}_D^{(h)}(W)$ and $p_n \in \mathcal{Q}_{IC}^{(h)}(W)$, where the superscript (h) indicates that the corresponding notion is considered with respect to $\mathcal{H}_{V,h}$. Since

$$\{\|g_n\|_{D,W}^{(h)}\}^2 = \delta_{g_n}^{(h)}(W) = \int_W h^2\, d\delta_{g_n} \le (\sup_W h)^2\, \delta_{g_n}(U') \to 0$$

$$(n \to \infty),$$

Lemma 6.3 implies that

$$\|u_n\|_{D,W}^{(h)} \to 0, \qquad \|p_n\|_{D,W}^{(h)} \to 0 \qquad (n \to \infty).$$

(Note that $\|\cdot\|_{D,W}^{(h)} = \|\cdot\|_{E,W}^{(h)}$, since $\sigma^{(h)}(1) = 0$.) By Corollary 6.7, $u_n + c_n \to 0$ locally uniformly on W for some sequence of constants $\{c_n\}$. On the other hand, by Lemma 4.12,

$$\int_W p_n^2\, d\lambda \le (\sup_W G_W^{(h)\lambda})\, I_W^{(h)}(\sigma^{(h)}(p_n)) = (\sup_W G_W^{(h)\lambda})\{\|p_n\|_{D,W}^{(h)}\}^2 .$$

$$\to 0 \quad (n \to \infty)$$

for any $\lambda \in \mathcal{M}_C^+(U')$. It follows that $\int_K (g_n+c_n)^2 \, d\lambda \to 0$ $(n \to \infty)$ for any compact set K in W and for any $\lambda \in \mathcal{M}_C^+(U')$. Then we see that $\{c_n\}$ can be chosen independent of W. Furthermore, since $\int_K g_n^2 \, d\nu \to 0$, $c_n \to 0$ if Supp $\nu \cap V \neq \emptyset$. Hence $\int_K g_n^2 \, d\lambda \to 0$ $(n \to \infty)$ for any compact set K in V if Supp $\nu \cap V \neq \emptyset$. Since any point of U' has a neighborhood V as above, Supp $\nu \cap U' \neq \emptyset$ and since U' is connected, it follows that $\int_{U'} g_n^2 \, d\lambda \to 0$ for any $\lambda \in \mathcal{M}_{C,o}^+(U')$, in particular, $\int_{U'} g_n^2 \, d\mu \to 0$.
Hence $\int_U f_n^2 \, d\mu \to 0$.

Corollary 7.2. Let U be a domain such that $\sigma(1) \neq 0$ on U and let $\mu \in \mathcal{M}_{C,o}^+(U)$. Then there is $M = M(\mu) > 0$ such that

$$\int_U f^2 \, d\mu \leq M \, \|f\|_{E,U}^2$$

for all $f \in \mathcal{D}_E(U)$.

Corollary 7.3. Let U be any domain. If $\mu, \nu \in \mathcal{M}_{C,o}^+(U)$ and $\mu \neq 0$, $\nu \neq 0$, then $\|\cdot\|_{D,\mu,U}$ and $\|\cdot\|_{D,\nu,U}$ are equivalent norms on $\mathcal{D}_D(U)$.

For a domain U in X, we denote by $\mathcal{D}_{D,o}(U)$ the closure of $\mathcal{D}_o(U) = \mathcal{D}(U) \cap \mathcal{C}_o(U)$ in $\mathcal{D}_D(U)$ with respect to the norm $\|\cdot\|_{D,\mu,U}$ for some $\mu \in \mathcal{M}_{C,o}^+(U)$. By Corollary 7.3 , we see that the space $\mathcal{D}_{D,o}(U)$ is independent of the choice of μ. In case $\sigma(1) \neq 0$ on U, we denote by $\mathcal{D}_{E,o}(U)$ the closure of $\mathcal{D}_o(U)$ in $\mathcal{D}_E(U)$ with respect to the norm $\|\cdot\|_{E,U}$; in case $\sigma(1) = 0$ on U, let $\mathcal{D}_{E,o}(U) = \mathcal{D}_{D,o}(U)$. By Corollary 7.2, $\mathcal{D}_{E,o}(U) \subset \mathcal{D}_{D,o}(U)$ in general.

By Lemma 7.2 , we see that $\mathcal{R}_o(U)$ is dense in $\mathcal{D}_{D,o}(U)$ (resp. in $\mathcal{D}_{E,o}(U)$ in case $\sigma(1) \neq 0$ on U) with respect to $\|\cdot\|_{D,\mu,U}$ (resp. $\|\cdot\|_{E,U}$). Furthermore, we have

Lemma 7.4. For any $f \in \mathcal{D}_{D,o}(U)$ (resp. $f \in \mathcal{D}_{E,o}(U)$ in case $\sigma(1) \neq 0$ on U), there exist $g_n \in \mathcal{R}_o(U)$, $n = 1,2,\ldots$, such that $\|f - g_n\|_{D,\mu,U} \to 0$ (resp. $\|f - g_n\|_{E,U} \to 0$) and $g_n \to f$ δ_f-a.e. on U as $n \to \infty$.

Proof. Since $\mathcal{R}_o(U)$ is dense in $\mathcal{D}_{D,o}(U)$ (resp. $\mathcal{D}_{E,o}(U)$), we can find $f_n \in \mathcal{R}_o(U)$, $n = 1,2,\ldots$, such that

$$\|f_n - f\|_{D,\mu,U} \to 0 \qquad \text{(resp. } \|f_n - f\|_{E,U} \to 0) \quad (n \to \infty)$$

with some $\mu \in \mathcal{M}^+_{C,o}(U)$. Since

$$\delta_{f_n} \le |f_n|\ |\sigma(f_n)| + \tfrac{1}{2}\sigma(f_n^2)^- + \tfrac{1}{2}f_n^2\sigma(1)^-$$

and f_n, $f_n^2 \in \mathcal{R}_o(U)$, we see that $\delta_{f_n} \in \mathcal{M}^+_{C,o}(U)$ for each n. Hence, by Lemma 7.3,

$$\int_U (f_n - f)^2\, d\delta_{f_m} \to 0 \qquad (n \to \infty)$$

for each m. Hence, by a diagonal method, we can choose a subsequence $\{f_{n_j}\}$ of $\{f_n\}$ such that

$$f_{n_j} \to f \qquad \delta_{f_m}\text{-a.e.} \qquad \text{for all } m.$$

Now, let $A = \{x \in U \mid f_{n_j}(x) \not\to f(x)\}$. Then $\delta_{f_m}(A) = 0$ for each m. Since $\delta_{f-f_m}(U) \to 0$ $(m \to \infty)$, it follows that $\delta_f(A) = 0$. Hence $f_{n_j} \to f$ δ_f-a.e., so that it is enough to take $g_j = f_{n_j}$.

Lemma 7.5. Let $f_n, f \in \mathcal{D}(U)$, $n = 1,2,\ldots$, and suppose $\delta_{f_n - f}(U) \to 0$ and $f_n \to f$ δ_f-a.e. on U $(n \to \infty)$. Then $\delta_{|f_n| - |f|}(U) \to 0$ $(n \to \infty)$.

Proof. Let

$A_n^+ = \{x\in U \mid f_n(x) > 0\}$, $A_n^- = \{x\in U \mid f_n(x) < 0\}$,

$A_n^o = \{x\in U \mid f_n(x) = 0\}$,

$A^+ = \{x\in U \mid f(x) > 0\}$, $A^- = \{x\in U \mid f(x) < 0\}$,

$A^o = \{x\in U \mid f(x) = 0\}$.

By Proposition 7.4, $\delta_{|f_n|}(A_n^o) = \delta_{f_n}(A_n^o) = 0$ and $\delta_{|f|}(A^o) =$

$= \delta_f(A^o) = 0$.

Hence,

$$\delta_{|f_n|-|f|}\{(A_n^+\cup A_n^o)\cap A^+\} = \delta_{f_n-f}\{(A_n^+\cup A_n^o)\cap A^+\},$$

$$\delta_{|f_n|-|f|}\{(A_n^-\cup A_n^o)\cap A^-\} = \delta_{f_n-f}\{(A_n^-\cup A_n^o)\cap A^-\},$$

$$\delta_{|f_n|-|f|}(A^o) = \delta_{|f_n|}(A^o) = \delta_{f_n}(A^o) = \delta_{f_n-f}(A^o),$$

$$\delta_{|f_n|-|f|}(A_n^-\cap A^+) = \delta_{f_n+f}(A_n^-\cap A^+)$$

$$\leq 8\ \delta_f(A_n^-\cap A^+) + 2\delta_{f_n-f}(A_n^-\cap A^+)$$

$$\delta_{|f_n|-|f|}(A_n^+\cap A^-) = \delta_{f_n+f}(A_n^+\cap A^-)$$

$$\leq 8\ \delta_f(A_n^+\cap A^-) + 2\delta_{f_n-f}(A_n^+\cap A^-).$$

Therefore,

$$\delta_{|f_n|-|f|}(U) \leq 2\delta_{f_n-f}(U) + 8\ \delta_f\{(A_n^-\cap A^+)\cup(A_n^+\cap A^-)\}.$$

Let $E_m = \bigcup_{n=m}^{\infty}\{(A_n^-\cap A^+)\cup(A_n^+\cap A^-)\}$ and $E = \bigcap_{m=1}^{\infty} E_m$. If $x\in E$, then $f_n(x) \nrightarrow f(x)$. Since $f_n \to f$ δ_f-a.e., it follows that $\delta_f(E) = 0$. Since $\{E_m\}$ is decreasing, $\delta_f(E_m) \to 0$ $(m \to \infty)$, and hence

$$\delta_f\{(A_n^- \cap A^+) \cup (A_n^+ \cap A^-)\} \to 0 \qquad (n \to \infty).$$

Therefore, $\delta_{|f_n|-|f|}(U) \to 0 \qquad (n \to \infty)$.

Proposition 7.5. $\mathcal{D}_{D,o}(U)$ and $\mathcal{D}_{E,o}(U)$ are closed under max. and min. operations.

Proof. Let $f \in \mathcal{D}_{D,o}(U)$ (resp. $\mathcal{D}_{E,o}(U)$). By Lemma 7.4, there exist $g_n \in \mathcal{R}_o(U)$, $n = 1,2,\ldots$, such that $\|f - g_n\|_{D,\mu,U} \to 0$ (resp. $\|f - g_n\|_{E,U} \to 0$) and $g_n \to f$ δ_f-a.e. on U ($n \to \infty$). Then, by Lemma 7.5, $\delta_{|f|-|g_n|}(U) \to 0$ ($n \to \infty$), which implies that $\|\,|f| - |g_n|\,\|_{D,\mu,U} \to 0$ (resp. $\|\,|f| - |g_n|\,\|_{E,U} \to 0$) ($n \to \infty$). Since $|g_n| \in \mathcal{R}_o(U) \subset \mathcal{D}_o(U)$, this means that $|f| \in \mathcal{D}_{D,o}(U)$ (resp. $\mathcal{D}_{E,o}(U)$).

7-4. Another extension of Green's formula

Lemma 7.6. If $f \in \mathcal{D}_{D,o}(U)$ (resp. $\mathcal{D}_{E,o}(U)$), then we can find $f_n \in \mathcal{R}_o(U)$ such that $|f_n| \leq |f|$ and $\|f_n - f\|_{D,\mu,U} \to 0$ (resp. $\|f_n - f\|_{E,U} \to 0$) ($n \to \infty$).

Proof. Let $f^+ = \max(f,0)$ and $f^- = \max(-f,0)$. Using Lemma 7.5 (cf. the proof of Proposition 7.5), we can find $\varphi_n, \psi_n \in \mathcal{R}_o^+(U)$, $n=1,2,\ldots$, such that

$$\|\varphi_n - f^+\| \to 0 \qquad \text{and} \qquad \|\psi_n - f^-\| \to 0 \qquad (n \to \infty),$$

where $\|\cdot\| = \|\cdot\|_{D,\mu,U}$ (resp. $= \|\cdot\|_{E,U}$). Let

$$\tilde{\varphi}_n = \min(\varphi_n, f^+) \quad \text{and} \quad \tilde{\psi}_n = \min(\psi_n, f^-).$$

Since $f^+ - \tilde{\varphi}_n = \max(f^+ - \varphi_n, 0)$, Corollary 7.1, (a) implies that

$$\delta_{f^+-\tilde{\varphi}_n}(U) \leq \delta_{f^+-\varphi_n}(U).$$

Obviously, $|f^+ - \tilde{\varphi}_n| \le |f^+ - \varphi_n|$. Hence

$$\|f^+ - \tilde{\varphi}_n\| \le \|f^+ - \varphi_n\| \to 0 \qquad (n \to \infty).$$

Similarly,

$$\|f^- - \tilde{\psi}_n\| \le \|f^- - \psi_n\| \to 0 \qquad (n \to \infty).$$

Thus, if we put $f_n = \tilde{\varphi}_n - \tilde{\psi}_n$, then $|f_n| \le |f|$ and $\|f - f_n\| \to 0 \quad (n \to \infty)$.

Theorem 7.3. (Green's formula) Let U be a domain in X. If $f \in \mathcal{R}_D(U)$, $g \in \mathcal{D}_{D,o}(U)$, $\int_U |fg|\, d|\sigma(1)| < +\infty$ and $\int_U |g|\, d|\sigma(f)| < +\infty$, then

$$\delta_{[f,g]}(U) + \int_U fg\, d\sigma(1) = \int_U g\, d\sigma(f).$$

In particular, if $u \in \mathcal{H}_E(U)$ and $g \in \mathcal{D}_{E,o}(U)$, then

$$\delta_{[u,g]}(U) + \int_U ug\, d\sigma(1) = 0.$$

Proof. By the previous lemma, we can find $g_n \in \mathcal{R}_o(U)$, $n = 1,2,\ldots$, such that $\|g_n - g\|_{D,\mu,U} \to 0$ $(n \to \infty)$ and $|g_n| \le |g|$ for all n. Since $|\sigma(1)| \in \mathcal{M}_C^+(U)$ and $|\sigma(f)| \in \mathcal{M}_C^+(U)$, Lemma 7.3 implies

$$\int_K (g_n - g)^2\, d|\sigma(1)| \to 0 \quad \text{and} \quad \int_K (g_n - g)^2\, d|\sigma(f)| \to 0$$

for any compact set K in U. Hence, choosing a subsequence if necessary we may assume that $g_n \to g$ $(|\sigma(1)| + |\sigma(f)|)$-a.e. on U. By Theorem 5.3 (or Theorem 7.1),

$$\delta_{[f,g_n]}(U) + \int_U fg_n\, d\sigma(1) = \int_U g_n\, d\sigma(f).$$

Since $\delta_{g_n - g}(U) \to 0$, $\delta_{[f,g_n]}(U) \to \delta_{[f,g]}(U)$ $(n \to \infty)$.

Since $|g_n| \leq |g|$, Lebesgue's convergence theorem implies $\int_U fg_n \, d\sigma(1) \to \int_U fg \, d\sigma(1)$ and $\int_U g_n \, d\sigma(f) \to \int_U g \, d\sigma(f)$. Hence we obtain the first formula of the theorem. To obtain the second formula, it is enough to remark that $u \in \mathcal{H}_E(U)$ implies $\sigma(u) = 0$, and $u \in \mathcal{H}_E(U)$ and $g \in \mathcal{D}_{E,o}(U)$ imply $\int_U |ug| \, d|\sigma(1)| < +\infty$.

Remark 7.1. If U is a PB-domain, then by Proposition 6.6, (c), $\mathcal{Q}_{IC}(U) \subset \mathcal{Q}_{EC}(U) \subset \mathcal{D}_{E,o}(U)$. Thus Theorem 7.4 includes Theorem 5.2 and Proposition 6.7.

7-5. Royden decomposition

Lemma 7.7. If U is a PB-domain, then

$$(2\beta_U - 1)^{-1} \|f\|_{E,U}^2 \leq \delta_f(U) + \int_U f^2 \, d\sigma(1) \leq \|f\|_{E,U}^2$$

for any $f \in \mathcal{D}_{E,o}(U)$.

Proof. This is immediate from Proposition 6.7 and the definition of $\mathcal{D}_{E,o}(U)$.

Lemma 7.8. Let U be a PB-domain and $f = u + g$ with $u \in \mathcal{H}_E(U)$ and $g \in \mathcal{D}_{E,o}(U)$. Then

$$\|u\|_{E,U} \leq M(\beta_U) \|f\|_{E,U}$$

and

$$\|g\|_{E,U} \leq (2\beta_U - 1) \|f\|_{E,U},$$

where $M(\cdot)$ is the function given in Lemma 6.3.

Proof. Choose $g_n \in \mathcal{R}_o(U)$, $n = 1,2,\ldots$, such that $\|g_n - g\|_{E,U} \to 0$ and put $f_n = u + g_n$. Then $\|f - f_n\|_{E,U} \to 0$. Hence our lemma immediately follows from Lemma 6.3.

Lemma 7.9. Let U be a PB-domain and $\mu \in \mathcal{M}^+_{C,o}(U)$. Then

$$\int_U g^2 \, d\mu \leq (\sup G_U^\mu) \|g\|^2_{E,U}$$

for all $g \in \mathcal{D}_{E,o}(U)$.

Proof. This is immediate from Lemmas 4.12 and 7.3.

Lemma 7.10. Let $U \in \mathcal{O}_X$ and V be a PC-domain such that $\overline{V} \subset U$. Then, for any $f \in \mathcal{D}(U)$,

$$f|V \in \mathcal{H}_E(V) + \mathcal{D}_{E,o}(V).$$

Furthermore, if $f|V = u + g$ with $u \in \mathcal{H}_E(V)$ and $g \in \mathcal{D}_{E,o}(V)$, then

(a) there is a sequence $\{\varphi_n\}$ in $\mathcal{R}_o(V)$ such that $\|\varphi_n - g\|_{E,V} \to 0$ and $\varphi_n \to g$ locally uniformly on V,

(b) $\beta_V \min(\inf_V f, 0) \leq u \leq \beta_V \max(\sup_V f, 0)$.

Proof. By definition, we can find $f_n \in \mathcal{R}(U)$ such that $\delta_{f_n - f}(V) \to 0$ and $f_n \to f$ uniformly on V $(n \to \infty)$. Let $f_n = u_n + g_n$ on V with $u_n \in \mathcal{H}_E(V)$ and $g_n \in \mathcal{Q}_{IC}(V)$ (Lemma 5.7). Then

(7.8) $$\beta_V \min(\inf_V f_n, 0) \leq u_n \leq \beta_V \max(\sup_V f_n, 0),$$

and $\{u_n\}$ is uniformly convergent on V, since $g_n \in \mathcal{Q}_{IC}(V)$. Since $\|f_n - f_m\|_{E,V} \to 0$ $(n,m \to \infty)$, Lemma 6.3 implies that $\|u_n - u_m\|_{E,V} \to 0$ $(n,m \to \infty)$. Hence, by Theorem 6.5 , we see that $u = \lim_{n\to\infty} u_n$ belongs to $\mathcal{H}_E(V)$ and $\|u_n - u\|_{E,V} \to 0$ $(n\to\infty)$. Let $g = f - u$ on V. Then $g_n \to g$ uniformly on V and $\|g_n - g\|_{E,V} \to 0$ $(n \to \infty)$. Now, let $\{W_n\}$ be an exhaustion of V. Then, by Proposition 6.6 , (c), we can find $\varphi_n \in \mathcal{R}_o(V)$ such that $\|\varphi_n - g_n\|_{E,V} \leq 1/n$ and $\sup_{W_n} |\varphi_n - g_n| < 1/n$ for each n. Then,

$\|\varphi_n - g\|_{E,V} \to 0$ and $\varphi_n \to g$ locally uniformly on V, which also implies that $g \in \mathcal{D}_{E,o}(V)$. Since $f_n \to f$ and $u_n \to u$ both uniformly on V, (7.8) implies the assertion (b).

Theorem 7.4. If U is a PB-domain, then

$$\mathcal{D}_E(U) = \mathcal{H}_E(U) \oplus \mathcal{D}_{E,o}(U) \qquad \text{(Royden decomposition).}$$

Furthermore,

(a) for any $g \in \mathcal{D}_{E,o}(U)$, there is a sequence $\{\varphi_n\}$ in $\mathcal{R}_o(U)$ such that $\|\varphi_n - g\|_{E,U} \to 0$ and $\varphi_n \to g$ locally uniformly on U;

(b) if $f = u + g$ with $u \in \mathcal{H}_E(U)$ and $g \in \mathcal{D}_{E,o}(U)$, then

$$\beta_U \min(\inf_U f, 0) \leq u \leq \beta_U \max(\sup_V f, 0),$$

in particular, $f \geq 0$ implies $u \geq 0$ and $|f| \leq M$ implies $|u| \leq \beta_U M$.

Proof. If $f \in \mathcal{H}_E(U) \cap \mathcal{D}_{E,o}(U)$, then by Theorem 7.3,

$$\delta_f(U) + \int_U f^2 \, d\sigma(1) = 0.$$

Hence, Lemma 7.7 implies that $\|f\|_{E,U} = 0$, so that $f = 0$ in case $\sigma(1) \not\equiv 0$ on U and $f = $ const. in case $\sigma(1) = 0$ on U; in the latter case, we also see that $f = 0$, using Lemma 7.9. Therefore, $\mathcal{H}_E(U) \cap \mathcal{D}_{E,o}(U) = \{0\}$, so that the vector sum $\mathcal{H}_E(U) + \mathcal{D}_{E,o}(U)$ is direct. Obviously, $\mathcal{H}_E(U) + \mathcal{D}_{E,o}(U) \subset \mathcal{D}_E(U)$. To prove the converse inclusion, take any $f \in \mathcal{D}_E(U)$. Let $\{W_n\}$ be an exhaustion of U such that each W_n is a domain. By the previous lemma,

$$f|W_n = u_n + g_n \quad \text{with } u_n \in \mathcal{H}_E(W_n),\ g_n \in \mathcal{D}_{E,o}(W_n),$$

and furthermore, we can choose $\varphi_n \in \mathcal{R}_o(W_n)$ such that

$$\|\varphi_n - g_n\|_{E,W_n} < \frac{1}{n} \quad \text{and} \quad \sup_{W_{n-1}} |\varphi_n - g_n| < \frac{1}{n}$$

for each n ($W_o = 0$). Also, we have

(7.9) $\quad \beta_U \min(\inf_U f, 0) \leq u_n \leq \beta_U \max(\sup_U f, 0)$ on W_n.

Each φ_n can be regarded as an element of $\mathcal{R}_o(U)$. By Lemma 7.8, $\{\|g_n\|_{E,W_n}\}$ is bounded, and hence $\{\|\varphi_n\|_{E,U}\}$ is bounded. Then, in exactly the same way as in the proof of Proposition 6.6, we see that $\|\varphi_n - \varphi_m\|_{E,U} \to 0$ $(n,m \to \infty)$, and hence $\|g_n - g_m\|_{E,W_n} \to 0$ $(m > n \to \infty)$ and $\|u_n - u_m\|_{E,W_n} \to 0$ $(m > n \to \infty)$, which implies that there is $u \in \mathcal{H}_E(U)$ such that $\|u_n - u\|_{E,W_m} \to 0$ $(n \to \infty)$ for each m and $u_n \to u$ locally uniformly on U, by virtue of Theorem 6.5. Let $g = f - u$. Then $g_n \to g$ locally uniformly on U, and hence $\varphi_n \to g$ locally uniformly on U. Also, as in the proof of Proposition 6.6 , we see that $\|\varphi_n - g\|_{E,U} \to 0$ $(n \to \infty)$. Hence $g \in \mathcal{D}_{E,o}(U)$, so that $f \in \mathcal{H}_E(U) + \mathcal{D}_{E,o}(U)$. Furthermore, if $f \in \mathcal{D}_{E,o}(U)$, then $u = 0$ in the above arguments, since $\mathcal{H}_E(U) \cap \mathcal{D}_{E,o}(U) = \{0\}$. Hence we have also obtained the assertion (a). Assertion (b) follows from (7.9).

§8. Royden boundary

8-1. Royden algebra (cf. [29] for the classical case)

For $U \in \mathcal{O}_X$, we consider the space

$$\mathcal{D}_{DB}(U) = \{f \in \mathcal{D}_D(U) \mid f\text{: bounded on } U\}$$

and the norm

$$\|f\|_{DB,U} = \delta_f(U)^{1/2} + \sup_U |f|.$$

$\mathcal{D}_{DB}(U)$ is a normed space with respect to this norm. By Proposition 7.4, $\mathcal{D}_{DB}(U)$ is closed under max. and min. operations and $\| |f| \|_{DB,U} = \|f\|_{DB,U}$.

<u>Theorem 8.1.</u> $\mathcal{D}_{DB}(U)$ is a Banach algebra with respect to the norm $\|\cdot\|_{DB,U}$.

<u>Proof.</u> If $f, g \in \mathcal{D}_{DB}(U)$, then fg is bounded and by Proposition 7.1, (b), we have

$$\delta_{fg}(U) = \int_U f^2\, d\delta_g + 2\int_U fg d\delta_{[f,g]} + \int_U g^2\, d\delta_f$$

$$\leq (\sup_U |f|)^2 \delta_g(U) + 2(\sup_U |f|)\cdot(\sup_U |g|)\delta_f(U)^{1/2} \cdot \delta_g(U)^{1/2} + (\sup_U |g|)^2 \delta_f(U)$$

$$= \{(\sup_U |f|)\delta_g(U)^{1/2} + (\sup_U |g|)\delta_f(U)^{1/2}\}^2 < +\infty.$$

Hence $fg \in \mathcal{D}_{DB}(U)$ and

$$\|fg\|_{DB,U} \leq (\sup_U |f|)\delta_g(U)^{1/2} + (\sup_U |g|)\delta_f(U)^{1/2} + (\sup_U |f|)(\sup_U |g|)$$

$$\leq \|f\|_{DB,U}\, \|g\|_{DB,U}.$$

Therefore, $\mathcal{D}_{DB}(U)$ is a normed algebra. If $\{f_n\}$ is a Cauchy sequence in $\mathcal{D}_{DB}(U)$ with respect to $\|\cdot\|_{DB,U}$, then obviously it is a Cauchy sequence in $\mathcal{D}(U)$ with respect to $\{P_K\}_{K:\text{compact}\subset U}$ (cf. 7-1) and $\{f_n\}$ is uniformly convergent on U. Hence $f = \lim_{n\to\infty} f_n$ belongs to $\mathcal{D}(U)$ and is bounded on U. For each compact set K in U,

$$\delta_f(K) = \lim_{n\to\infty} \delta_{f_n}(K) \leq \liminf_{n\to\infty} \delta_{f_n}(U).$$

Hence

$$\delta_f(U) \le \liminf_{n\to\infty} \delta_{f_n}(U) < +\infty.$$

Therefore, $f \in \mathcal{D}_{DB}(U)$. Furthermore,

$$\delta_{f-f_n}(K) = \lim_{m\to\infty} \delta_{f_m-f_n}(K) \le \liminf_{m\to\infty} \delta_{f_m-f_n}(U)$$

for any compact set K in U, so that

$$\delta_{f-f_n}(U) \le \liminf_{m\to\infty} \delta_{f_m-f_n}(U) \to 0 \qquad (n \to \infty).$$

It follows that $\|f_n - f\|_{DB,U} \to 0$ $(n \to \infty)$, which means that $\mathcal{D}_{DB}(U)$ is complete. Therefore, $\mathcal{D}_{DB}(U)$ is a Banach algebra.

In the classical case, $\mathcal{D}_{DB}(U)$ is called the Royden algebra on U (cf. [29]).

Next, we consider a subspace

$$\mathcal{D}_{DB,\Delta}(U) = \left\{ f \in \mathcal{D}_{DB}(U) \;\middle|\; \begin{array}{l} \text{there are } f_n \in \mathcal{D}_o(U) \text{ such that } \{f_n\} \\ \text{is uniformly bounded on } U,\ f_n \to f \\ \text{locally uniformly on } U \text{ and } \delta_{f_n-f}(U) \to 0 \end{array} \right\}.$$

This is a linear subspace of $\mathcal{D}_{DB}(U)$. It is easy to see that

$$\mathcal{D}_{DB,\Delta}(U) \subset \mathcal{D}_{D,o}(U) \cap \mathcal{D}_{DB}(U).$$

Proposition 8.1. $\mathcal{D}_{DB,\Delta}(U)$ is an ideal in $\mathcal{D}_{DB}(U)$, i.e., if $f \in \mathcal{D}_{DB,\Delta}(U)$ and $g \in \mathcal{D}_{DB}(U)$, then $fg \in \mathcal{D}_{DB,\Delta}(U)$.

Proof. Choose $f_n \in \mathcal{D}_o(U)$ such that $\{f_n\}$ is uniformly bounded on U, $f_n \to f$ locally uniformly on U and $\delta_{f_n-f}(U) \to 0$ $(n \to \infty)$. Then $f_n g \in \mathcal{D}_o(U)$ for each n, $\{f_n g\}$ is uniformly bounded on U and $f_n g \to fg$ locally uniformly on U.

Furthermore,

$$\delta_{f_n g - fg} = g^2 \delta_{f_n - f} + 2(f_n - f) g \delta_{[f_n - f, g]} + (f_n - f)^2 \delta_g .$$

Since g is bounded and $\delta_{f_n - f}(U) \to 0$, $\int_U g^2 \, d\delta_{f_n - f} \to 0$ $(n \to \infty)$. Since $\{f_n\}$ is uniformly bounded on U, say $|f_n| \le M$ on U for all n, f, g are bounded on U and $\delta_g(U) < +\infty$, we have

$$\left| \int_U (f_n - f) g \, d\delta_{[f_n - f, g]} \right| \le (M + \sup_U |f|)(\sup_U |g|) \delta_{f_n - f}(U)^{1/2} \cdot \delta_g(U)^{1/2}$$

$$\to 0 \quad (n \to \infty).$$

Also, since $\{(f_n - f)^2\}$ is uniformly bounded on U and $\delta_g(U) < +\infty$, Lebesgue's convergence theorem implies that $\int_U (f_n - f)^2 \, d\delta_g \to 0$ $(n \to \infty)$. Hence $\delta_{f_n g - fg}(U) \to 0$ $(n \to \infty)$, so that $fg \in \mathcal{D}_{DB,\Delta}(U)$.

<u>Corollary 8.1.</u> $1 \in \mathcal{D}_{DB,\Delta}(U)$ if and only if $\mathcal{D}_{DB,\Delta}(U) = \mathcal{D}_{DB}(U)$.

We say that U is <u>parabolic</u> if $1 \in \mathcal{D}_{DB,\Delta}(U)$, or equivalently, $\mathcal{D}_{DB,\Delta}(U) = \mathcal{D}_{DB}(U)$.

<u>Proposition 8.2.</u> If $f \in \mathcal{D}_{BD}(U)$ and $f \ge \alpha$ on U for some constant $\alpha > 0$, then $f^{1/2} \in \mathcal{D}_{BD}(U)$.

<u>Proof.</u> By Proposition 7.3, $f^{1/2} \in \mathcal{D}(U)$ and

$$\delta_{f^{1/2}} = \frac{1}{4f} \delta_f \le \frac{1}{4\alpha} \delta_f .$$

Hence $f^{1/2} \in \mathcal{D}_D(U)$. Obviously, $f^{1/2}$ is bounded on U.

<u>Proposition 8.3.</u> $\mathcal{D}_{DB,\Delta}(U)$ is closed under max. and min. operations. In fact, more generally, if $f \in \mathcal{D}_{DB,\Delta}(U)$, $g \in \mathcal{D}(U)$ and $0 \le g \le |f|$ on U, then there exists a sequence

$\{g_n\}$ in $\mathcal{D}_o(U)$ such that $g_n \geq 0$ on U for each n, $\{g_n\}$ is uniformly bounded, $g_n \to g$ locally uniformly on U and $\delta_{g_n - g}(U) \to 0$ $(n \to \infty)$, so that $g \in \mathcal{D}_{DB,\Delta}(U)$.

Proof. Choose $\{f_n\}$ in $\mathcal{D}_o(U)$ such that $\{f_n\}$ is uniformly bounded, $f_n \to f$ locally uniformly on U and $\delta_{f_n - f}(U) \to 0$. Let $g_n = \min(g, |f_n|)$ for each n. Then $g_n \in \mathcal{D}_o(U)$, $g_n \geq 0$, $\{g_n\}$ is uniformly bounded and $g_n \to g$ locally uniformly on U. Since $g_n - g = \frac{1}{2}\{|f_n| - |f| + (|g - |f_n|| - |g - |f| |)\}$, by repeated use of Lemma 7.5 , we see that $\delta_{g_n - g}(U) \to 0$ $(n \to \infty)$.

Proposition 8.4. If U is a PB-domain such that $|\sigma(1)|(U) < +\infty$, then $\mathcal{D}_{DB}(U) \subset \mathcal{D}_E(U)$, $\mathcal{H}_{BD}(U) = \mathcal{H}_E(U) \cap \mathcal{D}_{DB}(U) = \mathcal{H}_{BE}(U)$, $\mathcal{D}_{DB,\Delta}(U) = \mathcal{D}_{E,o}(U) \cap \mathcal{D}_{DB}(U)$ and

$$\mathcal{D}_{DB}(U) = \mathcal{H}_{BD}(U) \oplus \mathcal{D}_{DB,\Delta}(U).$$

Proof. The first two relations and $\mathcal{D}_{DB,\Delta}(U) \subset \mathcal{D}_{E,o}(U) \cap \mathcal{D}_{DB}(U)$ are obvious. Let $f \in \mathcal{D}_{E,o}(U) \cap \mathcal{D}_{DB}(U)$ and let $|f| \leq M$ on U. By Theorem 7.4 , there is a sequence $\{\varphi_n\}$ in $\mathcal{R}_o(U)$ such that $\|\varphi_n - f\|_{E,U} \to 0$ and $\varphi_n \to f$ locally uniformly on U. Put

$$f_n = \max(-M, \min(\varphi_n, M))$$

for each n. Then $f_n \in \mathcal{R}_o(U) \subset \mathcal{D}_o(U)$ for each n, $\{f_n\}$ is uniformly bounded on U and $f_n \to f$ locally uniformly on U. By repeated use of Lemma 7.5 , we also see that $\delta_{f_n - f}(U) \to 0$ $(n \to \infty)$. Hence $f \in \mathcal{D}_{DB,\Delta}(U)$. Now the last equality is a consequence of the Royden decomposition in Theorem 7.4.

<u>Corollary 8.2.</u> If U is a PB-domain, $|\sigma(1)|(U) < +\infty$ and $\mathcal{H}_{BD}(U) \neq \{0\}$, then U is not parabolic. In particular, if U is a P-domain and $\sigma(1) = 0$ on U, then U is not parabolic.

<u>Proof.</u> If U is a PB-domain, $|\sigma(1)|(U) < +\infty$ and $\mathcal{H}_{BD}(U) \neq \{0\}$, then by Proposition 8.4, $\mathcal{D}_{DB}(U) \neq \mathcal{D}_{DB,\Delta}(U)$, so that U is not parabolic. If U is a P-domain and $\sigma(1) = 0$ on U, then $1 \in \mathcal{H}_{BD}(U)$, so that $\mathcal{H}_{BD}(U) \neq \{0\}$. Hence U is not parabolic.

<u>8-2. Royden boundary</u> (cf. [29] for the classical case)

By a compactification of a locally compact Hausdorff space X, we mean a compact Hausdorff space X^* containing X as a dense open subset. There is a unique (up to homeomorphism) compactification X^* of X such that every $f \in \mathcal{D}_{DB}(X)$ can be continuously extended to X^* and the class of extended functions separates points of $X^*\backslash X$. This compactification is called the <u>Royden compactification</u> of X and $\Gamma = X^*\backslash X$ is called the <u>Royden boundary</u> of X.

<u>Remark 8.1.</u> The Royden compactification X^* is realized as the set of all characters on $\mathcal{D}_{DB}(X)$ with the w*-topology. Here, a character means a multiplicative linear functional κ on the algebra $\mathcal{D}_{DB}(X)$ such that $\kappa(1) = 1$. Such a character is a positive linear functional on $\mathcal{D}_{DB}(X)$, so that it belongs to the dual Banach space $\mathcal{D}_{DB}(X)'$ and $\|\kappa\| \leq 1$. The set $\tilde{X}$ of all characters on $\mathcal{D}_{DB}(X)$ is a w*-closed subset of the dual ball $\{\ell \in \mathcal{D}_{DB}(X)' \mid \|\ell\| \leq 1\}$, and hence w*-compact. By the imbedding $x \to \kappa_x \in \tilde{X} : \kappa_x(f) = f(x)$, we may regard X as an open dense subset of $\tilde{X}$ with respect to the w*-topology.

The extension of $f \in \mathcal{D}_{DB}(X)$ to X^* is again denoted by f.
Let $\mathcal{D}(\Gamma) = \{f|\Gamma \mid f \in \mathcal{D}_{DB}(X)\}$. Then, by the Stone-Weierstrass theorem, $\mathcal{D}(\Gamma)$ is dense in $\mathcal{C}(\Gamma)$ with the sup-norm.
Next, let

$$\Gamma_h = \{\xi \in \Gamma \mid f(\xi) = 0 \quad \text{for all } f \in \mathcal{D}_{DB,\Delta}(X)\}.$$

Then, Γ_h is a compact subset of Γ. Γ_h is called the <u>Royden harmonic boundary</u>. By Corollary 8.1 and the next lemma, $\Gamma_h = \emptyset$ if and only if X is parabolic.

<u>Lemma 8.1.</u> Let F be a closed subset of X^* such that $F \cap \Gamma_h = \emptyset$. Then there exists $f \in \mathcal{D}_{DB,\Delta}(X)$ such that $f = 1$ on F and $0 \leq f \leq 1$ on X^*.

<u>Proof.</u> Let $x \in F$. Since $x \notin \Gamma_h$, there exists $f_x \in \mathcal{D}_{DB,\Delta}(X)$ such that $f_x(x) \neq 0$. By Proposition 8.3, we may assume $f_x \geq 0$ on X. Then $f_x(x) > 0$. Since f_x is continuous on X^*, there is an open neighborhood V_x of x in X^* such that $f_x > 0$ on V_x. Since F is compact, we can choose a finite number of points $x_1, \ldots,$ $x_n \in F$ such that $V_{x_1} \cup \ldots \cup V_{x_n} \supset F$. Let $g = f_{x_1} + \ldots + f_{x_n}$. Then $g \in \mathcal{D}_{DB,\Delta}(X)$, $g \geq 0$ on X and $g > 0$ on F. Put $\alpha = \inf_F g$. Then $\alpha > 0$. Hence $f = \min(1, g/\alpha)$ is the required function in view of Proposition 8.3.

<u>Theorem 8.2.</u> (Minimum and maximum principles) Suppose X itself is a non-parabolic PB-domain and $|\sigma(1)|(X) < +\infty$. Let $u \in \mathcal{H}_{BD}(X)$. If $u \geq 0$ on Γ_h, then $u \geq 0$ on X; if $|u| \leq M$ on Γ_h, then $|u| \leq \beta_X M$ on X.

<u>Proof.</u> By Proposition 8.4,

$$1 = u_X + g_X \qquad \text{with } u_X \in \mathcal{H}_{BD}(X) \text{ and } g_X \in \mathcal{D}_{DB,\Delta}(X),$$

and by Theorem 7.4 , (b), $0 \leq u_X \leq \beta_X$. First, suppose $u \geq 0$ on Γ_h. Let $\varepsilon > 0$ and consider the set

$$K_\varepsilon = \{x \in X^* \mid u(x) + \varepsilon \leq 0\}.$$

Then K_ε is a closed set in X^* and $K_\varepsilon \cap \Gamma_h = \emptyset$. By the previous lemma, there exists $f \in \mathcal{D}_{DB,\Delta}(X)$ such that $f = 1$ on K_ε and $0 \leq f \leq 1$ on X^*. Since u is bounded, $u \geq -M$ for some $M \geq 0$. Then, $u + \varepsilon + Mf \geq 0$ on X^*.

Since

$$u + \varepsilon + Mf = (u + \varepsilon u_X) + (\varepsilon g_X + Mf),$$

$u + \varepsilon u_X \in \mathcal{H}_{BD}(X)$ and $\varepsilon g_X + Mf \in \mathcal{D}_{DB,\Delta}(X)$, Theorem 7.4 , (b) implies that $u + \varepsilon u_X \geq 0$. Since $\varepsilon > 0$ is arbitrary, it follows that $u \geq 0$ on X.

Next, suppose $|u| \leq M$ on Γ_h. Then, $Mu_X - u \geq 0$ on Γ_h, since $g_X = 0$ on Γ_h. Hence, by the above result, $u \leq Mu_X \leq \beta_X M$ on X. Similarly, we see that $-u \leq \beta_X M$.

Corollary 8.3. Suppose X is a PB-domain and $|\sigma(1)|(X) < +\infty$. Then,

$$\mathcal{D}_{DB,\Delta}(X) = \{f \in \mathcal{D}_{DB}(X) \mid f = 0 \text{ on } \Gamma_h\}.$$

Proof. Let $\mathcal{D}_1 = \{f \in \mathcal{D}_{DB}(X) \mid f = 0 \text{ on } \Gamma_h\}$. By the definition of Γ_h, $\mathcal{D}_{DB,\Delta}(X) \subset \mathcal{D}_1$. Let $f \in \mathcal{D}_1$ and $f = u + g$ with $u \in \mathcal{H}_{BD}(X)$ and $g \in \mathcal{D}_{DB,\Delta}(X)$. Then $u = f - g = 0$ on Γ_h. Hence by the above theorem, $u = 0$ on X, so that $f = g \in \mathcal{D}_{DB,\Delta}(X)$.

Hereafter, we assume that X is a non-parabolic PB-domain and $|\sigma(1)|(X) < +\infty$. Let

$$\mathcal{D}(\Gamma_h) = \{f|\Gamma_h \mid f \in \mathcal{D}_{DB}(X)\}.$$

This is a linear subspace of $\mathcal{C}(\Gamma_h)$ and is dense in $\mathcal{C}(\Gamma_h)$ with respect to the sup-norm.

Corollary 8.4. Given $\varphi \in \mathcal{D}(\Gamma_h)$, there is a unique $u_\varphi \in \mathcal{H}_{BD}(X)$ such that $u_\varphi|\Gamma_h = \varphi$. Furthermore, $\varphi \geq 0$ implies $u_\varphi \geq 0$; $|\varphi| \leq M$ implies $|u_\varphi| \leq \beta_X M$.

Proof. Choose $f \in \mathcal{D}_{DB}(X)$ such that $f|\Gamma_h = \varphi$ and let $f = u + g$ with $u \in \mathcal{H}_{BD}(X)$ and $g \in \mathcal{D}_{DB,\Delta}(X)$. Then $u|\Gamma_h = \varphi$. Uniqueness and the last assertion of the corollary immediately follow from Theorem 8.2.

According to this corollary, for each $x \in X$, the mapping

$$\varphi \mapsto u_\varphi(x)$$

is a bounded positive linear functional on $\mathcal{D}(\Gamma_h)$. Since $\mathcal{D}(\Gamma_h)$ is dense in $\mathbf{C}(\Gamma_h)$, it can be continuously extended to be a positive linear functional on $\mathbf{C}(\Gamma_h)$. Therefore, there exists a non-negative measure μ_x on Γ_h such that

$$\int_{\Gamma_h} \varphi \, d\mu_x = u_\varphi(x) \qquad \text{for all } \varphi \in \mathcal{D}(\Gamma_h).$$

By Harnack's inequality (Proposition 6.1 or Theorem 6.1), we see that μ_x and $\mu_{x'}$ are equivalent on Γ_h for any $x, x' \in X$.

<u>Theorem 8.3.</u> (Dirichlet problem with respect to the Royden harmonic boundary) For any $\varphi \in \mathbf{C}(\Gamma_h)$,

$$u_\varphi(x) = \int_{\Gamma_h} \varphi \, d\mu_x, \qquad x \in X$$

defines a bounded harmonic function u_φ on X such that

$$u_\varphi(x) \to \varphi(\xi) \quad \text{as } x \to \xi, \; x \in X$$

for all $\xi \in \Gamma_h$.

<u>Proof.</u> Since $\mathcal{D}(\Gamma_h)$ is dense in $\mathbf{C}(\Gamma_h)$, we can choose $\varphi_n \in \mathcal{D}(\Gamma_h)$ such that $\varphi_n \to \varphi$ uniformly on Γ_h. By Corollary 8.4 , we see that $\{u_{\varphi_n}\}$ is uniformly convergent on X. Since

$$u_\varphi(x) = \int_{\Gamma_h} \varphi \, d\mu_x = \lim_{n\to\infty} \int_{\Gamma_h} \varphi_n \, d\mu_x = \lim_{n\to\infty} u_{\varphi_n}(x)$$

for each $x \in X$, $u_\varphi \in \mathcal{H}(X)$. Since u_{φ_n} is continuous on X^*, $u_{\varphi_n}|\Gamma_h = \varphi_n$ and $\varphi_n \to \varphi$ uniformly on Γ_h, we see that u_φ extended by φ on Γ_h is continuous on $X \cup \Gamma_h$. Hence $u_\varphi(x) \to \varphi(\xi)$ as $x \to \xi$ for all $\xi \in \Gamma_h$.

8-3. Normal derivatives and Neumann problems (cf. [19], [20] for the classical case)

In this subsection, we still assume that X is a non-parabolic PB-domain and $|\sigma(1)|(X) < +\infty$. Let

$$\mathcal{R}_F(X) = \{f\in \mathcal{R}_E(X) \mid |\sigma(f)|(X) < +\infty\}.$$

$\mathcal{R}_F(X)$ is a linear space containing constants. Furhtermore, $\mathcal{H}_E(X) \subset \mathcal{R}_F(X)$.

Given $f\in \mathcal{R}_F(X)$, we define a linear functional $\ell_f : \mathcal{D}(\Gamma_h) \to \mathbb{R}$ by

(8.1) $\qquad \ell_f(\varphi) = \delta_{[f,u_\varphi]}(X) + \int_X fu_\varphi\, d\sigma(1) - \int_X u_\varphi\, d\sigma(f)$

for $\varphi\in \mathcal{D}(\Gamma_h)$. We say that f has a normal derivative on Γ_h (or on Γ) if ℓ_f is a bounded linear funtional on $\mathcal{D}(\Gamma_h)$ with respect to the sup-norm, i.e., if there exists $c_f \geq 0$ such that

$$|\ell_f(\varphi)| \leq c_f \sup_{\Gamma_h} |\varphi| \quad \text{for all } \varphi\in \mathcal{D}(\Gamma_h).$$

In this case, there is a unique signed measure ν_f on Γ_h such that

(8.2) $\qquad \ell_f(\varphi) = \int_{\Gamma_h} \varphi\, d\nu_f \qquad$ for all $\varphi\in \mathcal{D}(\Gamma_h)$.

This measure ν_f is called the normal derivative of $f\in \mathcal{R}_F(X)$. If f = const., then $\ell_f(\varphi) = 0$ for all $\varphi\in \mathcal{D}(\Gamma_h)$, so that f has normal derivative 0 on Γ_h. If $f_1, f_2\in \mathcal{R}_F(X)$ have normal derivatives on Γ_h and if λ_1,λ_2 are constants, then $\lambda_1 f_1 + \lambda_2 f_2$ has a normal derivative on Γ_h and $\nu_{\lambda_1 f_1+\lambda_2 f_2} = \lambda_1\nu_{f_1} + \lambda_2\nu_{f_2}$.

Proposition 8.5. (Generalized Green's formula) If $f\in \mathcal{R}_F(X)$ has normal derivative ν_f on Γ_h, then

$$\delta_{[f,g]}(X) + \int_X fg\, d\sigma(1) - \int_X g\, d\sigma(f) = \int_{\Gamma_h} g\, d\nu_f$$

for all $g\in \mathcal{D}_{DB}(X)$.

Proof. Let $g = u + g_o$ with $u \in \mathcal{H}_{BD}(X)$ and $g_o \in \mathcal{D}_{DB,\Delta}(X)$. Let $\varphi = g|\Gamma_h$. Then $u = u_\varphi$. Hence, by (8.1) and (8.2),

$$\delta_{[f,u]}(X) + \int_X fu\, d\sigma(1) - \int_X u\, d\sigma(f) = \int_{\Gamma_h} g\, d\nu_f.$$

On the other hand, since $\mathcal{D}_{DB,\Delta}(X) \subset \mathcal{D}_{D,o}(X)$, Theorem 7.3 implies that

$$\delta_{[f,g_o]}(X) + \int_X fg_o\, d\sigma(1) - \int_X g_o\, d\sigma(f) = 0.$$

Adding these two equations, we obtain the required formula.

Corollary 8.5. If $f, g \in \mathcal{R}_F(X)$ are bounded on X and have normal derivatives ν_f, ν_g, respectively on Γ_h, then

$$\int_X f\, d\sigma(g) - \int_X g\, d\sigma(f) = \int_{\Gamma_h} g\, d\nu_f - \int_{\Gamma_h} f\, d\nu_g.$$

In particular, if $u,v \in \mathcal{H}_{BD}(X)$ have normal derivatives ν_u, ν_v, respectively, on Γ_h, then

$$\int_{\Gamma_h} u\, d\nu_v = \int_{\Gamma_h} v\, d\nu_u \quad \text{and} \quad \int_{\Gamma_h} d\nu_u = \int_X u\, d\sigma(1).$$

Proposition 8.6. If $g \in \mathcal{Q}_{EC}(X)$ and $|\sigma(g)|(X) < +\infty$, then g has normal derivative ν_g on Γ_h. If, in particular $\sigma(g) \geq 0$ (i.e., $g \in \mathcal{P}_{EC}(X)$), then $\nu_g \leq 0$.

Proof. By Proposition 6.7 (or Theorem 7.3),

$$\delta_{[g,u_\varphi]}(X) + \int_X gu_\varphi\, d\sigma(1) = 0$$

for any $\varphi \in \mathcal{D}(\Gamma_h)$. Hence, $\ell_g(\varphi) = -\int_X u_\varphi\, d\sigma(g)$. Thus, by Corollary 8.4,

$$|\ell_g(\varphi)| \leq \beta_X\{|\sigma(g)|(X)\}(\sup_{\Gamma_h} |\varphi|)$$

for all $\varphi \in \mathcal{D}(\Gamma_h)$. Hence g has a normal derivative on Γ_h.

If $\sigma(g) \geq 0$, then again by Corollary 8.4, $\ell_g(\varphi) \leq 0$ for any $\varphi \in \mathcal{D}^+(\Gamma_h)$, so that $\nu_g \leq 0$.

Finally, we consider Neumann problems with respect to the Royden harmonic boundary Γ_h.

Lemma 8.2. For each $x \in X$, there is $M_x > 0$ such that

$$\int_{\Gamma_h} \varphi^2 \, d\mu_x \leq M_x \| u_\varphi \|_1^2$$

for all $\varphi \in \mathcal{D}(\Gamma_h)$, where

$$\|u\|_1^2 = \begin{cases} \|u\|_{E,X}^2 & \text{if } \sigma(1) \not\equiv 0 \text{ on } X \\ \|u\|_{D,X}^2 + u(x)^2 & \text{if } \sigma(1) = 0 \text{ on } X. \end{cases}$$

Proof. First, we note that $\mathcal{H}_{BD}(X) = \mathcal{H}_{BE}(X)$ under our assumption. Thus, if $u \in \mathcal{H}_{BD}(X)$, then, since $\sigma(u^2) = -2\delta_u - u^2\sigma(1)$, we have

(8.3) $$|\sigma(u^2)|(X) \leq 2 \|u\|_{E,X}^2 .$$

It follows that $u^2 \in \mathcal{H}(X) + \mathcal{Q}_C(X)$. As is easily seen, $u^2 \in \mathcal{R}_E(X)$. Hence, by Proposition 6.6 , (b),

$$u^2 = h + g \qquad \text{with } h \in \mathcal{H}_E(X) \text{ and } g \in \mathcal{Q}_{EC}(X).$$

Since u^2 is bounded, so is h; hence $h \in \mathcal{H}_{BD}(X)$. By Proposition 6.6 ,(c), $\mathcal{Q}_{EC}(X) \subset \mathcal{D}_{E,o}(X)$. Since g is also bounded, Proposition 8.4 implies that $g \in \mathcal{D}_{DB,\Delta}(X)$.

Now, let $\varphi \in \mathcal{D}(\Gamma_h)$. Then, as we have seen above,

$$u_\varphi^2 = h_\varphi + g_\varphi \quad \text{with } h_\varphi \in \mathcal{H}_{BD}(X) \text{ and } g_\varphi \in \mathcal{Q}_{EC}(X) \cap \mathcal{D}_{DB,\Delta}(X).$$

Since $h_\varphi|\Gamma_h = u_\varphi^2|\Gamma_h = \varphi^2$, $h_\varphi = u_{\varphi^2}$. Hence,

(8.4) $$u_\varphi^2 = u_{\varphi^2} + g_\varphi \quad \text{with} \quad g_\varphi = G_X^\nu \,, \; \nu = \sigma(u_\varphi^2).$$

Suppose now the assertion of the lemma is not true for some $x \in X$. Then we could find $\varphi_n \in \mathcal{D}(\Gamma_h)$, $n = 1,2,\ldots$, such that

$$u_{\varphi_n^2}(x) = 1 \quad \text{and} \quad \|u_{\varphi_n}\|_1^2 < 2^{-n},\; n = 1,2,\ldots\;.$$

Put $\nu_n = \sigma(u_{\varphi_n}^2)$ for each n. By (8.3), we see that $|\nu_n|(X) < 2^{1-n}$. Hence $\mu = \Sigma_{n=1}^\infty |\nu_n|$ is a non-negative measure on X with finite total mass, so that G_X^μ is a potential. It follows that $\Sigma_{n=1}^\infty G_X^{|\nu_n|}$ is finite on a dense set in X, which implies that $g_{\varphi_n} \to 0$ on a dense set in X. On the other hand, $\|u_{\varphi_n}\|_1 \to 0$ implies that $u_{\varphi_n} \to 0$ locally uniformly on X (Corollaries 6.6 and 6.7). Hence by (8.4) $u_{\varphi_n^2} \to 0$ on a dense set in X. Since $u_{\varphi_n^2} \geq 0$ by Corollary 8.4 , Harnack's inequality (Proposition 6.1 or Theorem 6.1) implies that $u_{\varphi_n^2} \to 0$ locally uniformly on X which contradicts our assumption that $u_{\varphi_n^2}(x) = 1$ for all n. Hence the lemma is proved.

<u>Theorem 8.4.</u> (Neumann problem) Suppose $\sigma(1) \geq 0$ on X. Fix $x_o \in X$ and let $\gamma \in L^2(\Gamma_h;\mu_{x_o})$. If $\sigma(1) = 0$ on X, assume furthermore $\int \gamma \, d\mu_{x_o} = 0$. Then there exists $u \in \mathcal{H}_E(X)$ which has normal derivative $\nu_u = \gamma\mu_{x_o}$ on Γ_h; u is unique if $\sigma(1) \not\equiv 0$ on X and u is unique up to additive constants if $\sigma(1) = 0$ on X.

<u>Proof.</u> For φ, $\psi \in \mathcal{D}(\Gamma_h)$, let

(8.5)
$$\langle\varphi,\psi\rangle = \delta_{[u_\varphi,u_\psi]}(X) + \int_X u_\varphi u_\psi \, d\sigma(1) \qquad \text{if } \sigma(1) \neq 0,$$
$$\langle\varphi,\psi\rangle = \delta_{[u_\varphi,u_\psi]}(X) + u_\varphi(x_o)u_\psi(x_o) \qquad \text{if } \sigma(1) = 0.$$

Then, $\langle \cdot , \cdot \rangle$ is an inner product on $\mathcal{D}(\Gamma_h)$. Let $\hat{\mathcal{D}}$ be the completion of $\mathcal{D}(\Gamma_h)$ with respect to the corresponding norm $\|\varphi\| = \langle\varphi,\varphi\rangle^{1/2}$. Then $\hat{\mathcal{D}}$ is a Hilbert space. Since $\|u_\varphi\|_1 = \|\varphi\|$ for $\varphi \in \mathcal{D}(\Gamma_h)$, the mapping $\varphi \to u_\varphi$ can be extended to be an isometry from $\hat{\mathcal{D}}$ into $\mathcal{H}_E(X)$ with the norm $\|\cdot\|_1$ and (8.5) remains valid for any $\varphi,\psi \in \hat{\mathcal{D}}$. By Lemma 8.2 , we see that $\varphi \mapsto \int_{\Gamma_h} \varphi\gamma \, d\mu_{x_o}$ is a continuous linear form on $\mathcal{D}(\Gamma_h)$, so that it can be extended to be a continuous linear form on $\hat{\mathcal{D}}$. Hence there exists $\varphi_o \in \hat{\mathcal{D}}$ such that

(8.6)
$$\langle\varphi, \varphi_o\rangle = \int_{\Gamma_h} \varphi\gamma \, d\mu_{x_o}$$

for all $\varphi \in \mathcal{D}(\Gamma_h)$. If $\sigma(1) \neq 0$, then this means that $\gamma\mu_{x_o}$ is the normal derivative of u_{φ_o}, and hence u_{φ_o} is the required function. If $\sigma(1) = 0$, then (8.6) shows that

$$\delta_{[u_\varphi,u_{\varphi_o}]}(X) = \int_{\Gamma_h} \varphi\gamma \, d\mu_{x_o}$$

for all $\varphi \in \mathcal{D}(\Gamma_h)$ with $u_\varphi(x_o) = 0$. Thus, given $\varphi \in \mathcal{D}(\Gamma_h)$, applying this equation to $\psi = \varphi - u_\varphi(x_o)$, we obtain

$$\delta_{[u_\varphi,u_{\varphi_o}]}(X) = \int_{\Gamma_h} \{\varphi-u_\varphi(x_o)\}\gamma \, d\mu_{x_o} = \int_{\Gamma_h} \varphi\gamma \, d\mu_{x_o},$$

since $u_\psi = u_\varphi - u_\varphi(x_o)$ and $\int_{\Gamma_h} \gamma \, d\mu_{x_o} = 0$ by assumption. Hence $\gamma\mu_{x_o}$ is the normal derivative of u_{φ_o} in this case, too. If u_1, $u_2 \in \mathcal{H}_E(X)$ have the same normal derivative on Γ_h, then

(8.7) $$\delta_{[u_\varphi, u_1-u_2]}(X) + \int_X u_\varphi(u_1-u_2)\, d\sigma(1) = 0$$

for all $\varphi \in \mathcal{D}(\Gamma_h)$, and hence for all $\varphi \in \hat{\mathcal{D}}$. By Corollary 6.9, $\mathcal{H}_{BD}(X) = \mathcal{H}_{BE}(X)$ is dense in $\mathcal{H}_E(X)$ with respect to the norm $\|\cdot\|_1$. It follows that $\{u_\varphi \mid \varphi \in \hat{\mathcal{D}}\} = \mathcal{H}_E(X)$. Therefore, (8.7) implies that $\|u_1 - u_2\|_{E,X} = 0$, so that $u_1 = u_2$ if $\sigma(1) \neq 0$ and $u_1 = u_2$ + const. if $\sigma(1) = 0$.

The next proposition shows that there are many harmonic functions having normal derivatives on Γ_h:

Proposition 8.9. Suppose $\sigma(1) \geq 0$ on X. Then the set of all $u \in \mathcal{H}_E(X)$ which have normal derivatives on Γ_h is dense in $\mathcal{H}_E(X)$ with respect to the norm $\|\cdot\|_1$.

Proof. For $\gamma \in L^2(\Gamma_h;\mu_{x_o})$ ($\int_{\Gamma_h} \gamma\, d\mu_{x_o} = 0$ in case $\sigma(1) = 0$), let $\tilde{u}_\gamma$ be the solution of the Neumann problem $\nu_u = \gamma\mu_{x_o}$ (with $\tilde{u}_\gamma(x_o) = 0$ in case $\sigma(1) = 0$), and let

$$\mathcal{H}_N = \{\tilde{u}_\gamma \mid \gamma \in L^2(\Gamma_h:\mu_{x_o}) \text{ (and } \int \gamma\, d\mu_{x_o} = 0 \text{ in case } \sigma(1) = 0)\}$$

It is enough to show that $\mathcal{H}_N$ is dense in $\mathcal{H}_E(X)$ in case $\sigma(1) \neq 0$, dense in $\mathcal{H}_E^o(X) = \{u \in \mathcal{H}_E(X) \mid u(x_o) = 0\}$ in case $\sigma(1) = 0$. Since $\mathcal{H}_E(X) = \{u_\varphi \mid \varphi \in \hat{\mathcal{D}}\}$ as we showed at the end of the proof of the previous theorem, if $\mathcal{H}_N$ is not dense in $\mathcal{H}_E(X)$ (in case $\sigma(1) \neq 0$) or in $\mathcal{H}_E^o(X)$ (in case $\sigma(1) = 0$), then there would exist $\varphi \in \hat{\mathcal{D}}$ such that $u_\varphi \neq 0$ (and $u_\varphi(x_o) = 0$ in case $\sigma(1) = 0$) and

(8.8) $$\delta_{[u_\varphi, \tilde{u}_\gamma]}(X) + \int_X u_\varphi \tilde{u}_\gamma\, d\sigma(1) = 0$$

for all $\gamma \in L^2(\Gamma_h;\mu_{x_o})$ (with $\int \gamma\, d\mu_{x_o} = 0$ in case $\sigma(1) = 0$).

By Lemma 8.2 , we can find $\tilde{\varphi} \in L^2(\Gamma_h;\mu_{x_o})$ such that

$\int_{\Gamma_h}(\varphi_n-\tilde{\varphi})^2 d\mu_{x_o} \to 0$ for any sequence $\{\varphi_n\}$ in $\mathcal{D}(\Gamma_h)$ such that $\varphi_n \to \varphi$ in $\hat{\mathcal{D}}$. Since

$$\delta_{[u_{\varphi_n},\tilde{u}_\gamma]}(X) + \int_X u_{\varphi_n}\tilde{u}_\gamma \, d\sigma(1) = \int_{\Gamma_h} \varphi_n \gamma \, d\mu_{x_o},$$

(8.8) implies that

$$\int_{\Gamma_h} \tilde{\varphi}\gamma \, d\mu_{x_o} = 0$$

for all $\gamma \in L^2(\Gamma_h;\mu_{x_o})$ (with $\int \gamma \, d\mu_{x_o} = 0$ in case $\sigma(1) = 0$). Hence $\tilde{\varphi} = 0$, so that $\int_{\Gamma_h} \varphi_n^2 \, d\mu_{x_o} \to 0$ for the above $\{\varphi_n\}$. From Harnack's inequality, it follows that $\int_{\Gamma_h} \varphi_n^2 \, d\mu_x \to 0$ for any $x \in X$, and hence $u_{\varphi_n} \to 0$, i.e., $u_\varphi = 0$, a contradiction. Thus the proposition is proved.

Remark 8.2. Normal derivatives and Neumann problems can be considered for other kinds of ideal boundaries; see [19] and [20].

APPENDIX

Networks

Examples of harmonic spaces, whose base spaces are not manifolds, are supplied by networks. They are often good sources of counter examples (cf. [11, Exercises]).

(A-1) Definition of networks

Let A, B and C be at most countable sets and suppose $A \neq \emptyset$ and $B \cup C \neq \emptyset$. Let K be a function on $A \times B$ satisfying the following conditions:

(K.1) The range of K is $\{-1,0,1\}$;

(K.2) For each $b \in B$, $e(b) = \{a \in X \mid K(a,b) \neq 0\}$ consists of exactly two points a_1 and a_2 and $K(a_1,b)K(a_2,b) = -1$;

(K.3) For each $a \in A$, $B(a) = \{b \in B \mid K(a,b) \neq 0\}$ is a finite set;

(K.4) For each $a, a' \in A$ $(a \neq a')$, there are $a_1,\dots,a_k \in A$ and $b_1,\dots,b_{k+1} \in B$ such that $e(b_j) = \{a_{j-1},a_j\}$, $j = 1,\dots,k+1$, with $a_0 = a$ and $a_{k+1} = a'$.

We also consider a mapping F from C into A such that

(F) $F^{-1}(a)$ is a finite set (may be empty) for each $a \in A$.

For each $z \in B \cup C$, we consider a set S_z which is equivalent to the open unit interval $]0,1[$ by a bijection j_z, and let

$$X = A \cup \bigcup_{z \in B \cup C} S_z$$

be a disjoint union. We introduce a topology on X as follows: $U \subset X$ is open if and only if it satisfies the following (i)~(iv):

(i) $j_z(U \cap S_z)$ is open in $]0,1[$ for each $z \in B \cup C$,

(ii) if $a \in U \cap A$, $b \in B(a)$ and $K(a,b) = -1$, then $j_b(U \cap S_b)$ contains an interval of the form $]0,\varepsilon[$ $(\varepsilon > 0)$,

(iii) if $a\in U\cap A$, $b\in B(a)$ and $K(a,b) = 1$, then $j_b(U\cap S_b)$ contains an interval of the form $]1-\varepsilon',1[$ $(\varepsilon' > 0)$,

(iv) if $a\in U\cap A$ and $c\in F^{-1}(a)$, then $j_c(U\cap S_c)$ contains an interval of the form $]0,\varepsilon''[$ $(\varepsilon'' > 0)$.

Then, X is a connected, locally connected, locally compact Hausdorff space with a countable base. For each $b\in B$, j_b is extended to be a homeomorphism of $\overline{S}_b = S_b\cup e(b)$ onto $[0,1]$; the extended map will be again denoted by j_b. Note that $K(j_b(0),b) = -1$ and $K(j_b(1),b) = 1$. For each $c\in C$, j_c is extended to be a homeomorphism of $\overline{S}_c = S_c\cup\{F(c)\}$ onto $[0,1[$. We may consider an ideal point β_c for each $c\in C$ ($\beta_c\notin X$, $\beta_c \neq \beta_{c'}$ if $c \neq c'$) and define $j_c(\beta_c) = 1$. Then we can introduce a unique topology on $X^* = X\cup\{\beta_c \mid c\in C\}$ in such a way that its restriction to X is the same as the original topology of X and each j_c is a homeomorphism of $S_c^* = \overline{S}_c\cup\{\beta_c\}$ onto $[0,1]$. X^* is compact if and only if A is finite (in this case B and C are also finite); X is compact if and only if A is finite and $C = \emptyset$.

Next, we consider four functions $\gamma_{-1}, \gamma_1 : B \to [0,+\infty[$, $\zeta:C \to [0,+\infty[$ and $\rho:A \to R$ which satisfy

(A.1) $$\sum_{b\in B(a)} \gamma_{K(a,b)}(b) + \sum_{c\in F^{-1}(a)} \zeta(c) > 0 \qquad \text{for each } a\in A.$$

Then the configuration $\mathcal{N} = \{A,B,C,K,F,\gamma_{\pm 1},\zeta,\rho\}$ is called a <u>network</u>; it is called a finite (resp. infinite) network if A is a finite (resp. infinite) set. The topological space X defined above is called the base space corresponding to $\mathcal{N}$.

<u>(A-2) Harmonic structure associated with a network</u>

A subset U of some S_z $(z\in B\cup C)$ is called an interval if $j_z(U)$ is an interval. A function f on an interval $U\subset S_z$ is said to be linear on U if $f\circ j_z^{-1}$ is linear on $j_z(U)$. If f is linear on S_b, $b\in B$, then $\lim_{x\to a, x\in S_b} f(x)$ exists as a finite value for each $a\in e(b)$. This value will be denoted by $f(a;b)$. Similary, if f is linear on S_c, $c\in C$, the values $f(a;c)$ for $a = F(c)$ and $f(\beta_c) = \lim_{x \to \beta_c} f(x)$ are defined.

A domain U in X will be called proper if $U \cap S_z$ is an interval for each $z \in B \cup C$ such that $U \cap S_z \neq \emptyset$. For a proper domain U, let

$$\tilde{U} = \bigcup \{S_z \mid z \in B \cup C,\ S_z \cap U \neq \emptyset\} \cup (U \cap A).$$

Then $U \subset \tilde{U}$ and $\tilde{U}$ is a proper domain. If a function f on a proper domain U is linear on each $S_z \cap U \neq \emptyset$, then f has a linear extension on S_z for such z and obtain a function $\tilde{f}$ on $\tilde{U}$. This $\tilde{f}$ is called the linear extension of f.

Given a proper domain U in X, $u \in \mathcal{C}(U)$ is called harmonic on U (with respect to $\mathcal{N}$) if it satisfies the following two conditions:

(a) u is linear on $S_z \cap U$ for each $z \in B \cup C$ such that $S_z \cap U \neq \emptyset$;

(b) for each $a \in U \cap A$,

$$\text{(A.2)} \quad \sum_{b \in B(a)} K(a,b)\gamma_{K(a,b)}(b) \sum_{a' \in e(b)} K(a',b)\tilde{u}(a';b) + \sum_{c \in F^{-1}(a)} \zeta(c)\{u(a) - \tilde{u}(\beta_c)\} + \rho(a)u(a) = 0,$$

where $\tilde{u}$ is the linear extension of u on $\tilde{U}$.

If U is an open set, then $u \in \mathcal{C}(U)$ is called harmonic on U if it is harmonic on any proper subdomain of U. Let $\mathcal{H}(U)$ be the set of all harmonic functions on U. It is easy to see that $\mathcal{H}(U)$ is a linear subspace of $\mathcal{C}(U)$ and $\mathcal{H}$ is a sheaf. We now show

Proposition A.1. $(X, \mathcal{H})$ is a Bauer's harmonic space.

Proof. First, we verify Axiom 2 of Brelot. If $x \in X \setminus A$, then $x \in S_z$ for some $z \in B \cup C$. Let $t_o = j_z(x)$. Then $0 < t_o < 1$. Choose $\varepsilon > 0$ such that $]t_o-\varepsilon, t_o+\varepsilon[\subset]0,1[$. Then we easily see that $j_z^{-1}(]t_o-\varepsilon, t_o+\varepsilon[)$ is regular with respect to $\mathcal{H}$. Next, let $x = a \in A$. For $0 < \varepsilon < 1$, let

$$\text{(A.3)} \quad V_{a,\varepsilon} = \{a\} \cup \bigcup_{\substack{b \in B(a) \\ K(a,b)=-1}} j_b^{-1}(]0,\varepsilon[) \cup \bigcup_{\substack{b \in B(a) \\ K(a,b)=1}} j_b^{-1}(]1-\varepsilon,1[) \cup$$

$$\bigcup_{c\in F^{-1}(a)} j_c^{-1}(]0,\varepsilon[).$$

Note that

$$\partial V_{a,\varepsilon} = \{j_b^{-1}(\varepsilon) \mid b\in B(a), K(a,b) = -1\}$$

$$\cup \{j_b^{-1}(1-\varepsilon) \mid b\in B(a),\ K(a,b) = 1\} \cup \{j_c^{-1}(\varepsilon) \mid c \in F^{-1}(a)\}.$$

Given $\varphi\in \mathcal{C}(\partial V_{a,\varepsilon})$, put $\hat{\varphi}(z) = \varphi(j_z^{-1}(\varepsilon))$ if $z\in B(a)$, $K(a,z) = -1$, $\hat{\varphi}(z) = \varphi(j_z^{-1}(1-\varepsilon))$ if $z\in B(a)$, $K(a,z) = 1$ and $\hat{\varphi}(z) = \varphi(j_z^{-1}(\varepsilon))$ if $z\in F^{-1}(a)$.

If $u = H_{\varphi}^{V_{a,\varepsilon}}$ exists, then it is determined by $u(a)$ and φ, and $u(a)$ must satisfy

$$\sum_{b\in B(a)} \gamma_{K(a,b)}(b)\{u(a) - \hat{\varphi}(b)\} + \sum_{c\in F^{-1}(a)} \zeta(c)\{u(a) - \hat{\varphi}(c)\}$$

$$+ \varepsilon\rho(a)u(a) = 0$$

or

(A.4) $$\{\sum_{b\in B(a)} \gamma_{K(a,b)}(b) + \sum_{c\in F^{-1}(a)} \zeta(c) + \varepsilon\rho(a)\}u(a)$$

$$= \sum_{b\in B(a)} \gamma_{K(a,b)}(b)\hat{\varphi}(b) + \sum_{c\in F^{-1}(a)} \zeta(c)\hat{\varphi}(c).$$

By virtue of condition (A.1), there is $\varepsilon_a > 0$ such that if $0 < \varepsilon \leq \varepsilon_a$ then

(A.5) $$\sum_{b\in B(a)} \gamma_{K(a,b)}(b) + \sum_{c\in F^{-1}(a)} \zeta(c) + \varepsilon\rho(a) > 0.$$

Then the equation (A.4) has a unique solution $u(a)$, and furthermore, $\varphi \geq 0$ on $\partial V_{a,\varepsilon}$ implies $u(a) \geq 0$. Thus, $V_{a,\varepsilon}$, $0 < \varepsilon \leq \varepsilon_a$, are regular domains. Thus Axiom 2 of Brelot is satisfied.

Axiom (BC) is easily verified, since the limit of convergent sequence of linear functions is again linear and the sums appearing in (A.2) are all finite sums.

By (A.4), $H_1^{V_{a,\varepsilon}}(a) > 0$ for $0 < \varepsilon < \varepsilon_a$, which implies Axiom (P).

Finally, we verify Axiom (S). It is obvious that for a domain V of the form $V = j_z^{-1}(]t_1, t_2[)$ with $z \in B \cup C$ and $0 \leq t_1 < t_2 \leq 1$, $\mathcal{H}(V)$ separates points of V. Next, consider the domain $V = V_{a,\varepsilon}$ with $a \in A$ and $0 < \varepsilon < \varepsilon_a$. Let $\mathcal{U}_1(V)$ be the set of all continuous functions s on $\overline{V}$ such that s is linear on each $S_z \cap V$ with $z \in B(a) \cup F^{-1}(a)$ and

$$\sum_{b \in B(a)} \gamma_{K(a,b)}(b)\{s(a) - s(\xi_b)\} + \sum_{c \in F^{-1}(a)} \zeta(c)\{s(a) - s(\xi_c)\}$$

$$+ \varepsilon\rho(a)s(a) \geq 0,$$

where $\{\xi_z\} = \partial V \cap S_z$ for $z \in B(a) \cup F^{-1}(a)$. It is easy to see that each $s \in \mathcal{U}_1(V)$ is hyperharmonic on V. Let $s_o \in \mathcal{C}(\overline{V})$ be such that it is linear on each $S_z \cap V$ with $z \in B(a) \cap F^{-1}(a)$, $s_o(a) = 1$ and $s_o(\xi) = 0$ for all $\xi \in \partial V$. Then $s_o \in \mathcal{U}_1(V)$ by (A.5). Next, let φ be a function on ∂V such that $\varphi(\xi_z) > 0$ for each $z \in B(a) \cup F^{-1}(a)$ and $\varphi(\xi_z) \neq \varphi(\xi_{z'})$ if $z \neq z'$. Choose $s_1 \in \mathcal{C}(\overline{V})$ such that it is linear on each $S_z \cap V$ with $z \in B(a) \cap F^{-1}(a)$, $s_1(a) = 1$ and $s_1 = \varphi$ on ∂V. By choosing the values of φ sufficiently small, we see by (A.5) that $s_1 \in \mathcal{U}_1(V)$. It is easy to see that s_o and s_1 separate points of V. Therefore, Axiom (S) is satisfied and the proof of the proposition is completed.

Proposition A.2. $(X, \mathcal{H})$ is a Brelot's harmonic space if and only if $\gamma_{-1}, \gamma_1, \zeta$ are all strictly positive.

Proof. First, assume that γ_{-1}, γ_1 and ζ are all strictly positive. If U is an interval contained in some S_z, $z \in B \cup C$, and if $u_n \in \mathcal{H}(U)$, then $(u_n \circ j_z^{-1})(t) = a_n t + b_n$, $t \in]t_1, t_2[$. Hence, if $\{u_n\}$ is monotone increasing and $\{u_n(x_o)\}$ is bounded for some $x_o \in U$, then $\{a_n\}$ and $\{b_n\}$ converge to finite values, so that $u = \lim_{n\to\infty} u_n$ belongs to $\mathcal{H}(U)$. Next, suppose $U = V_{a,\varepsilon}$ in the notation used in the proof of the previous proposition, and let $u_n \in \mathcal{H}(U)$, $\{u_n\}$ be monotone increasing and $\{u_n(x_o)\}$ be bounded for some $x_o \in U$. In view of the above consideration on intervals, we may assume that $\{u_n(a)\}$ is bounded. Since

$$\{\sum_{b\in B(a)} \gamma_{K(a,b)}(b) + \sum_{c\in F^{-1}(a)} \zeta(c) + \varepsilon\rho(a)\}u_n(a) =$$

$$= \sum_{b\in B(a)} \gamma_{K(a,b)}(b)u_n(\xi_b) + \sum_{c\in F^{-1}(a)} \zeta(c)u_n(\xi_c)$$

and $\gamma_{\pm 1}$, ζ are strictly positive, we see that $\{u_n(\xi_b)\}$, $\{u_n(\xi_c)\}$ are bounded. Hence $u = \lim_{n\to\infty} u_n$ is harmonic on U. Now, it is easy to see that Axiom 3 of Brelot is satisfied.

Conversely, suppose one of γ_{-1}, γ_1, ζ is not strictly positive; e.g., suppose $\gamma_{-1}(b_o) = 0$ for $b_o \in B$. Choose $a \in A$ such that $K(a,b_o) = -1$ and consider the domain $V_{a,\varepsilon}$. Let $u_n(x) = 0$ if $x \in V_{a,\varepsilon} \setminus S_{b_o}$ and $u_n(x) = n j_{b_o}(x)$ if $x \in V_{a,\varepsilon} \cap S_{b_o}$. Then $u_n \in \mathcal{H}(V_{a,\varepsilon})$ for each n, $\{u_n\}$ is monotone increasing and $\{u_n(a)\}$ is bounded, while $u_n(x) \to +\infty$ if $x \in V_{a,\varepsilon} \cap S_{b_o}$.

Thus, Axiom 3 of Brelot fails to hold.

(A-3) Measure representation and gradient measures on networks

For an open set U in X and $f \in \mathcal{R}(U)$, if $U \cap S_z \neq \emptyset$, $z \in B \cup C$, then $(f\circ j_z^{-1})''$ in the distribution sense is a signed measure on $j_z(U \cap S_z)$. Let f_z'' denote the signed measure on $U \cap S_z$ which is the pull-back of $(f\circ j_z^{-1})''$ by j_z. Then, we see that $j_z(1-j_z)f_z''$ can be regarded as a signed measure on $U \cap \overline{S}_z$. Furthermore, if $a \in U$, $b \in B(a)$ (resp. $c \in F^{-1}(a)$) and $\gamma_{K(a,b)} \neq 0$ (resp. $\zeta(c) \neq 0$), then $f'(a;z) = \lim_{x\to a, x\in S_z \cap U} (f\circ j_b^{-1})'(j_b(x))$ exists and is finite, where $z = b$ (resp. $z = c$) and

$(f \circ j_z^{-1})'$ is the ordinary derivative which exists almost everywhere. Thus, we can define a signed measure $\sigma(f)$ on U by

$$\sigma(f) = - \sum_{z \in B \cup C,\ U \cap S_z \neq \emptyset} j_z(1-j_z)f''_z$$

$$+ \sum_{a \in A \cap U} \left\{ \sum_{b \in B(a)} \gamma_{K(a,b)}(b) f'(a;b) \right.$$

$$\left. + \sum_{c \in F^{-1}(a)} \zeta(c) f'(a;c) + \rho(a) f(a) \right\} \varepsilon_a,$$

where ε_a denotes the unit point mass at a. Then we see that σ defines a measure representation of $\mathcal{R}$. In the case where $(X, \mathcal{H})$ is a Brelot's harmonic space (i.e., the case where $\gamma_{\pm 1}$, ζ are all strictly positive), we may replace $j_z(1-j_z)f''_z$ in the first sum of the right hand side by f''_z .

The corresponding gradient measure is given by

$$\delta_{[f,g]} = \sum_{z \in B \cup C, U \cap S_z \neq \emptyset} j_z(1-j_z) j_z^*[(f \circ j_z^{-1})'(g \circ j_z^{-1})' dt],$$

for $f, g \in \mathcal{R}(U)$, where j_z^* denotes the pull-back by j_z. In particular, if $f \in \mathcal{R}(X)$ and f is linear on each S_z, $z \in B \cup C$, then

$$\delta_f(X) = \frac{1}{4} \left[\sum_{b \in B} \{ \sum_{a \in e(b)} K(a,b) f(a) \}^2 + \sum_{c \in C} \{ f(F(c)) - f(\beta_c) \}^2 \right].$$

REFERENCES

[1] H. Bauer, Harmonische Räume und ihre Potentialtheorie, Lecture Notes in Math. 22, Springer-Verlag, 1966.

[2] N. Boboc, C. Constantinescu and A. Cornea, On the Dirichlet problem in the axiomatic theory of harmonic functions, Nagoya Math. J. 23 (1963), 73-96.

[3] J.-M. Bony, Détermination des axiomatiques de théorie du potentiel dont les fonctions harmoniques sont différentiables, Ann. Inst. Fourier 17,1 (1967), 353-382.

[4] M. Brelot, Éléments de la théorie classique du potentiel, 4e éd., Centre Doc. Univ. Paris, 1969.

[5] M. Brelot, Étude et extensions du principe de Dirichlet, Ann. Inst. Fourier 5 (1955), 371-419.

[6] M. Brelot, Lectures on potential theory, Part IV, Tata Inst. F.R., 1960; Reissued 1967.

[7] M. Brelot, Axiomatique des fonctions harmoniques, Univ. Montréal, 1966.

[8] H. Cartan, Théorie du potentiel newtonien: énergie, capacité, suites de potentiels, Bull. Soc. Math. France 73 (1945), 74-106.

[9] H. Cartan, Théorie générale du balayage en potentiel newtonien, Ann. Univ. Grenoble 22 (1946), 221-280.

[10] C. Constantinescu and A. Cornea, Ideale Ränder Riemannscher Flächen, Springer-Verlag, 1963.

[11] C. Constantinescu and A. Cornea, Potential theory on harmonic spaces, Springer-Verlag, 1972.

[12] J. Deny and J.L. Lions, Les espaces du type de Beppo Levi, Ann. Inst. Fourier 5 (1955), 305-370.

[13] M. Glasner and M. Nakai, Riemannian manifolds with discontinuous metrics and the Dirichlet integral, Nagoya Math. J. 46 (1972), 1-48.

[14] W. Hansen, Cohomology in harmonic spaces, Seminar on potential theory II, Lecture Notes in Math. 226, 63-101, Springer-Verlag, 1971.

[15] L.L. Helms, Introduction to potential theory, Wiley-Interscience, 1969.

[16] R.-M. Hervé, Recherches axiomatiques sur la théorie des fonctions surharmoniques et du potentiel, Ann. Inst. Fourier 12 (1962), 415-571.

[17] K. Janssen, On the existence of a Green function for harmonic spaces, Math. Ann. 208 (1974), 295-303.

[18] P.A. Loeb and B. Walsh, The equivalence of Harnack's principle and Harnack's inequality in the axiomatic system of Brelot, Ann. Inst. Fourier 15,2 (1965), 597-600.

[19] F-Y. Maeda, Normal derivatives on an ideal boundary, J. Sci. Hiroshima Univ., Ser. A-I 28 (1964), 113-131.

[20] F-Y. Maeda, Boundary value problems for the equation $\Delta u - qu = 0$ with respect to an ideal boundary, Ibid. 32 (1968), 85-146.

[21] F-Y. Maeda, Harmonic and full-harmonic structures on a differentiable manifold, Ibid. 34 (1970), 271-312.

[22] F-Y. Maeda, Energy of functions on a self-adjoint harmonic space I, Hiroshima Math. J. 2 (1972), 313-337.

[23] F-Y. Maeda, Energy of functions on a self-adjoint harmonic space II, Ibid. 3 (1973), 37-60.

[24] F-Y. Maeda, Dirichlet integrals of functions on a self-adjoint harmonic space, Ibid. 4 (1974), 682-742.

[25] F-Y. Maeda, Dirichlet integrals of product of functions on a self-adjoint harmonic space, Ibid. 5 (1975), 197-214.

[26] F-Y. Maeda, Dirichlet integrals on general harmonic spaces, Ibid. 7 (1977), 119-133.

[27] F-Y. Maeda, Differential equations associated with harmonic spaces, Proceedings of the Colloquium on Complex Analysis, Joensuu 1978, Lecture Notes in Math. 747, 260-267, Springer-Verlag, 1979.

[28] M. Nakai, The space of Dirichlet-finite solutions of the equation $\Delta u = Pu$ on a Riemann surface, Nagoya Math. J. 18 (1961), 111-131.

[29] L. Sario and M. Nakai, Classification theory of Riemann surfaces, Springer-Verlag, 1970.

[30] G.L. Tautz, Zum Umkehrungsproblem bei elliptischen Differentialgleichungen I, II, Bemerkungen, Arch. Math. 3 (1952), 232-238, 239-250, 361-365.

[31] B. Walsh, Flux in axiomatic potential theory II : Duality, Ann. Inst. Fourier 19,2 (1969), 371-417.

[32] B. Walsh, Perturbation of harmonic structures and an index-zero theorem, Ibid. 20,1 (1970), 317-359.

[33] N.A. Watson, Green functions, potentials, and the Dirichlet problem for the heat equation, Proc. London Math. Soc. (3) 33 (1976), 251-298.

INDEX OF TERMINOLOGIES

INDEX OF SYMBOLS